Frontispiece
Centaur in a mist, winner of the 1994 Centenary Competition
photograph by Gloria Taylor

CENTAUR

- - - ≪ ≪ ≪ ◎ ≫ ≫ ≫ - - -

Commemoration of a Centenary

- - - ≪ ≪ ≪ ◎ ≫ ≫ ≫ - - -

Cross Channel Cargo Carrying to Club Charter Cruising

- - - ≪ ≪ ≪ ◎ ≫ ≫ ≫ - - -

**Compiled by Members of
The Thames Barge Sailing Club**

published 1995
by The Thames Barge Sailing Club
c/o
National Maritime Museum,
Greenwich,
London, SE10 9NF

ISBN   0 - 95 25 474 - 0 - 6

British Library Cataloguing in Publication Data
information available from the British Library

Printed in Great Britain by
Leighton Printing Co.
15 Palmer Place, Kilburn, London, N7 8DH

Subscribers

The following members have subscribed to the production of this Commemorative Book and wish Centaur success in the next 100 years

Pat Boss	Peter Boss
Gordon L.Brown	Maurice Butters
Alan Cordell	Brian Dawson
Dick Dawson	Jane Dawson
A.P.Dalby	Pam Elkins
Roy Emmins	Gwen Ives
Ken Ives	Tim Jepson
Tom Johnson	Glen Fenemore-Jones
Jane Kennedy	Ronnie Lavender
Simon Lavington	Barrie Lawrence
Frances Lewis	Tony Mackett
Helen Martyn	Roger Martyn
E.J.Matthews	Roger Newlyn
Denise Newlyn	Stan Owen
Diana Robbins	Walter Robbins
Pam Siggory	Ted Siggory
Doreen M. Stevenson	John L. Stevenson
Gloria Taylor	Val.Valentine Smith
Rita Valentine-Smith	Tony de Winton
David Wood	Elizabeth Wood
Charles Woodcraft	Marie Woodman
Steve Yates	Wendy Yates

Contents

PHOTOGRAPHS & OTHER ILLUSTRATIONS

Foreword

This book has been compiled to celebrate the centenary of the sailing barge CENTAUR by the Historical Literary and Photographic subcommittee of the Thames Barge Sailing Club under its Chairman, Elizabeth Wood and assisted by many members, including Tim Jepson, and Pat Boss. Throughout the 47 years of its existence the club has amassed a not inconsiderable archive of Bulletins, Journals, log books and a photographic collection and this is the basis of this Centenary book.

Some contributions are small, perhaps just a line or footnote, others, particularly the reports made by Peter Boss, both as Chair of the Barge Committee and as one of our skippers have grown over the years to be worthy of a separate publication. Some contributions to the literature of the club have already been the basis of earlier books, notably by Harvey Benham and Arthur Bennett. The Club is particularly grateful to other past contributors to our publications particularly Rod Minter, Felix Mallett, and Stan Yeates.

For the new material which this book has prompted we must especially thank Mrs.Annie Duke for her help and Frank Thompson, and John Glenn for their recollections of her last years in trade; Barry Pearce, for insight into her history and repair, John Prime, Eddie Smith, and Lindsey Simon must be thanked for details of her conversion and work as a charter barge; Mrs.Ruth Clarke for the light she has thrown on the builders and their family.

Photographs are credited where known and we are especially grateful to Nicholas Hardinge for his photographs taken in the last years in trade kindly printed for us by Phil Chatfield and to Pat Boss and Gloria Taylor who have taken so many photographs of CENTAUR and her crews in recent years.

So much of the life and work of CENTAUR has not been recorded and much which could have been written has been omitted through lack of space or time for thorough research. It must be regretted that so many records have been lost (or if written are unintelligible!) and the log books kept by the Duty Mates are so cryptic as not to describe in graphic detail what the crew would have discussed in the bar afterwards.

This book has been compiled on the Macintosh Performa 200 using ClarisWorks 0.1Bv3 including the 'apellling chequer' usually used for producing the Members Bulletin and the Hon. Secretary alone must accept responsibility for any errors and apologise for those omissions which, he hopes, others will point out.

D.G.Wood,
Hon. Secretary, T.B.S.C.
February 1995

Marjorie on the slipway ready for launch and Kimberley after her launch with the Gas works in the background

1. The new Channel Barge 1895

CENTAUR was launched from the yard of John and Herbert Cann at Bathside in Gashouse Creek, Harwich on Friday 15th February, 1895, one of the coldest winters on record. The Thames was frozen and shipping came to a standstill with barges and lighters sunk by ice flows. On 11th February the Medway froze over , the cement industry came to a halt, new barge launchings were delayed until the end of the month and at Chatham all outdoor work was at a standstill owing to the arctic weather. On 9th February 1895 *The Harwich & Dovercourt Free Press* reported that 'weather of arctic severity prevails in Essex, on one two occasions this week the thermometer has registered as much as from 25 to 30- degrees of frost.'

On the day of the launch *The Essex Weekly News* reported ice drifts in Brightlingsea Creek and arctic scenes on the River Crouch with that river impassible with three or four foot thick ice flows, some acres in extent. The cold weather continued for weeks and the *Essex Weekly News* reported on 29th February that there was a carnival on the ice at Chelmsford, and cycling and cricket on the Chelmer and Blackwater and opposite Howards ship building yard at Maldon there was a cricket match between teams from Maldon and Heybridge with much interest shown by spectators, albeit the strong east wind rendered their position somewhat uncomfortable while some 40 or 50 local watermen had their own primitive football contest on the ice. *The Harwich & Dovercourt Free Press* had an editorial on 16th February on "The Extraordinary Weather: The intensity of the great frost complicated in many cases by heavy weather and disastrous gales and by blinding snowstorms in others has continued over the whole area of these isles with a catalogue of deaths, illness and privation in consequence is a grievous one'. The Vicar of Harwich, H.L.Norden wrote to the paper that 'It is very evident if we have a continuance of the severe weather there will be great distress amongst the working classes. I should be glad to join with others in holding a public meeting to form a strong committee for preparing for any consequences that may arise'. The same issue has a note of the Annual General Meeting of the Harwich Barge Alliance Association the previous Thursday at the Great Eastern Hotel which following the recent death of J.H.Vaux was chaired by William Groom at which the talk was of the recent heavy gales, which, had fortunately resulted in no claims. It was in these conditions that the builders had laboured and launched the new barge.

The yard had been founded by their father, George Cann of Gt. Bentley in Essex who came to Harwich from Brightlingsea where he had been apprenticed. He went into partnership with a boat-builder and blockmaker as Parsons and Cann and built six ketch barges between 1872 and 1879 when the partnership was dissolved. Census returns for 1881 show George Cann as head of the household aged 43 being a Barge Builder and employing four men, living with his wife Sarah Ann aged 45 at 26 Daniel Street, in the parish of St.Nicholas, Harwich Their children living with them were John Garling Cann aged 16 who had been born at Brightlingsea, and then working as a bargebuilder, his sister Omelia, aged 14, and brothers Johnson Ernest aged 12, and Herbert James, born 4th December 1870 then aged 10, both described as Scholars and their cousin Kate Mastell aged 2, their other children George and Gertrude do not appear in the census return, George apparently having by that time left home.

The Cann Family
Herbert   Omelia   John   George   Gertrude   Johnson

The boat and barge building business obviously prospered and on 25th May 1889 George took on 14 year old Frank Ainger as apprentice at five shillings a week

14

increasing to a shilling a week each year until his 7th and final year. However on Tuesday 4th June, 1889 George was injured at the Harwich level crossing; he had returned to the shipyard from the auction of a partly burnt barge and went to help his employees unload timber from a railway truck. One of the trees slipped and fell on his chest crushing him against one of the trucks. He was quickly released and although he recovered slightly and bleeding from his mouth stopped, he died at home surrounded by his family including his brother in law, Mr.Garling who seems to have taken charge. An Inquest was held the following Thursday returning a verdict of accidental death and the jury asked the Coroner to write to the railway company and asked for a crane to be fixed for unloading heavy loads. *The Harwich & Dovercourt Newsman* for 8th June 1889 noted that the deceased was insured in the Railway Passengers Accidental Insurance Co through their agent at Ipswich, Mr.L.M.L.Sanders. The yard with a full order book was left in the hands of his son John aged 24 who signed a further deed to continue the apprenticeship of Frank Ainger on 14th February, 1890.

Herbert was still a minor, aged 18, and so could not legally be made a partner and the DOROTHY completed in 1889 is attributed in the maritime registers only to John but both brothers are named as builders of the boomie CARISBROOKE CASTLE launched in 1890. In November 1891 Daniel Street was renamed Albert Street and the four children continued to live together there with Johnson, then aged 22, described as a Grocer, as nominal head and John and Herbert described as shipwrights. Their business was listed as ships chandlers & saw mills & etc. Bathgate (Telegrams "Cann, Shipbuilders, Harwich") and blacksmiths, 1, West Street. and John and Herbert continued to trade as Cann Brothers building a fine fleet of spritsail barges, the last of which was LEOFLEDA built in 1915. They also built bawleys for the Harwich and Leigh fleets, yachts and motor craft, barge boats and sailing dinghies.

John made half models of new craft with his own hands and showed them to his brother for comment. These half models were not the varnished ones often seen hanging on directors walls, but large half sections on a board with ribbands to show the shape. Barry Pearce, former barge skipper, once owner of Walter Cook & Sons, Maldon, artist and historian was told of a large fully rigged model with a varnished hull had been made by John Cann as a present to the wife of one of the original owners but the model is said to have gone to relatives in America. It is thought that every barge launched was photographed by Herbert and that the glass plates still exist in the Harwich area but their whereabouts is unknown. Due to its position up a narrow creek by the gasworks all launches at the yard were made at high water by tipping the craft over the edge of the quay broadside with a tremendous splash. Photographs show MARJORIE on the slipway in 1899 and KIMBERLEY just after launch the next year with the gas holder in the background.

The brothers eventually moved away from Daniel Street, John and his wife Eliza to 'Stour View" Stour Street but their son George did not become involved in the business. Herbert moved first to 32 Fernlea Road and then to 'Alberta' Main Road He eventually became Vice-Consul for Holland, Belgium, Finland, and Latvia, and was also Lloyd's agent with an office at 41a Church Street. Herbert married his cousin, Kate Kerridge and they had eight children; she pased away on 29th July 1921 when, Basil, her youngest son was only eight. On 30th June 1923 he married Ellen Conway by whom he had one daughter. The yard closed around 1925 and John Cann died on 21st May, 1932. Herbert continued active through the 2nd World War until his death on 19th January, 1954 while his grocer brother survived until 28th June, 1964, so much may be gathered from official records.

At the time CENTAUR was built Rod Minter writes, "Cann's yard was producing some of the finest coasting barges on the east coast, and it was to this yard that Charles Stone, of the famous Mistley barging family of Stone Brothers went for a barge to be used both for trade and holidays. To this end she was exceptionally well fitted out with no expense spared. Her smartness and value as a trading barge was for a long time without equal on this part of the Essex coast. Down below she was given a mahogany finish to the cabin and red plush backing to the transom, transom slots for bottles, brackets to hold glasses, a water tank pump and the luxury of a stateroom. Other notable items included an eight day clock and mahogany weather glass. At some time she was also fitted with a large brass bound wheel, although this may well have been when she went to Brightlingsea, as much of the yacht equipment went on to the local barges".

Among the advertisements for drapery and notices of houses to let in the issue of *The Harwich and Dovercourt Newsman* for 16th February, 1895 there is a simple inch deep entry on the back page of local events:

| ꞌꞏꞏꞏꞏ⸗ꞏꞏg, Terrace, ——— achinery 31, Post ———. :porience | New Ship.—Messrs. J. and H. Cann, barge builders, dispatched from their building yard, Bathside, on Friday, the new barge Centaur, which they have built to the order of Captain Charles Stone, of Mistley. Her dimensions are as follows:—length 85·6, breadth 19·6, depth 6·6, and about 160 tons burden. Owing to the severe weather the usual trial trip was dispensed with. | The body Lady Mo F. Ochr brother, who was Bremesh⸗ the Cruf been encl and plac⸗ |

CENTAUR was registered in the Harwich Custom House as vessel 99460 on 18th February, No.1 in 1895 in Harwich and allotted the signal letters N.R.L.H. (later amended to M.Q.R.V.) The Transcript of her original registration shows that her registered gross tonnage was measured at 70.70 ton with additions of .69 ton for the cabin top and 4.88 for excess hatchways. CENTAUR's certificate showed both

FORM No. 10.
Signal Letters (if any)

Transcript of Register for Transmission
to Chief Registrar of Shipping.

M Q R V
N R L H

...ber of Ship	Name of Ship	No., Date, and Port of Registry
9460	Centaur	N 7 in 1895 Harwich

Whether British or Foreign Built	Whether a Sailing or Steam Ship; and if a Steam Ship, how propelled	Where Built	When Built	Name and Address of Builders
British	Sailing	Harwich	1895	J & H Cann, Harwich

			Feet	Tenths
Number of Decks	One	Length from fore part of stem, under the bowsprit, to the aft side of the head of the stern post	85	6
Number of Masts	Two			
Rigged	Spritsail	Main breadth to outside of plank	19	5
Stern	Square	Depth in hold from tonnage deck to ceiling at midships	6	2
Build	Carvel			
Galleries	None	Depth in hold from upper deck to ceiling at midships, in the case of three decks and upwards	—	—
Head	None			
Framework	Wood	Length of engine room, if any ..	—	—

20 FEB 95

Registry closed the 24th day of October, 1955. Vessel
converted into a lighter. Registry no longer required.
Advice received from beneficial owner. Concurrence
received from Registrar ... cancelled

PARTICULARS OF ENGINES (if any)

No. of Engines	Description	Whether British or Foreign made	When made	Name and Address of Makers	Diameter of Cylinders	Length of Stroke	No. of Horses power (combined)

PARTICULARS OF TONNAGE

		Tons			
GROSS TONNAGE:			DEDUCTIONS ALLOWED		
Under Tonnage Deck		60.70	On account of space required for propelling power		
Closed in spaces above the Tonnage Deck, if any			On account of spaces occupied by Seamen or Apprentices, and appropriated to their use, and kept free from Goods or Stores of every kind, not being the personal property of the Crew		
Space or spaces between Decks ..			These spaces are the following, viz.:—		
Poop			Lower Forecastle		6.04
Forecastle					
Round House					
Other closed in spaces, if any, as follows:					
Cabin Top		.69			
Space Hatchways		4.88	Cabin & Others		
			Master's Cabin	3.58	
		76.27	215.84	Seamen's Stores	2.00	8.30
Deductions, as per Contra ..		14.34	40.58	Last 3 Tons	1.71	
Registered Tonnage ..		61.93	175.26	Total Deductions ..		14.34

FOR CHANGE OF TONNAGE

| Name of Master | Charles Stone | Certificate of { Service No. { Competency No. | |

Names, Residence, and Description of the Owners, and Number of Sixty-fourth Shares held by each } viz.

Charles Stone, Bargeowner 40, Eliza Stone, wife of Charles Stone
Bargeowner, 16, both of Mistley; Mary Ann Barnett, wife of
Harry Barnett, Chartering Agent, of Forest Gate 4, all in the
County of Essex; and Caroline Underwood, wife of Harry
Underwood, Hay & Straw Merchant, of Brentford, County of
Middlesex, 4. Sixty four

Dated 18th February 1895   Registrar

NOTE.—If there are more Owners than one the Registrar is requested to distinguish the Managing Owner by the letters "M.O." against his name.

N.B.—To be addressed in an Envelope to the Chief Registrar of Shipping, Custom House, London.

(No. 361.)

Transcript of the Certificate of Registration,   Public Record Office file BT110,133 the imperial gross tonnage of 76.27 and a cubic metric equivalent of 215.84. From the gross figure there was deduced on account of spaces occupied by seaman and kept free from goods or stores comprising the lower forecastle 6.04; the masters cabin 3.58, b'swains' store 2.03, the sail room 1.01 and the chart locker .78, in all 14.34 ton or 40.58 cubic meters to arrive at her original registered tonnage of 61.93 (175.25 cubic metres). She was remeasured on 25th March 1919 with the slightly reduced registered tonnage of 59.76 ton to come below the 60 ton limit of her new trading port of Colchester. The original owners of her 64 shares were Charles Stone, bargeowner of 2, The Green, Mistley 40 shares ; Eliza Stone his wife 16 shares; Mary Ann Barnett wife of Harry Barnett, Chartering Agent of Forest Gate, 4 shares and Caroline Underwood, wife of Harry Underwood, Hay and Straw Merchant of Brentford in the County of Middlesex the remaining 4 shares.

On 5th July, 1895 Charles Stone signed the first of many forms entitled "Declaration of Half Yearly Agreement and Account of voyages and crew of the ship engaged in the home trade only". Chas. Stone, then aged 41, stated that he had previously served in the sailing barge PEGASUS of Rochester along with his mate 37 year old W.Smith while James Smith the boy aged 18 had previously served in the ketch barge ALICE WATTS of Harwich. The master received a share in the freight while the mate received £3.10.0. per week and the boy £1.

Reference No.	SIGNATURES OF CREW.	Age.	Nationality.*	If in the Reserve, No. of Commission or R.V.2.	Ship in which he last served, and Year of Discharge therefrom.		Date and Place of signing this Agreement.		In what capacity engaged, and if Master, Mate, or Engineer,† No. of Certificate.	Time at which he is to be on board.
					Year.	State Name and Official No. or Port she belonged to.	Date.	Place.		
	1	2	3	4	5	6	7	8	9	10
1	*Chas Stone* Master to sign first.	41	British	—	1895	*Pegasus Rochester*	27/7/95	Harwich	Master.	
2	*W. Smith*	37	Do	—	Do	*Pegasus Rochester*	– o –	– . .	*mate* *done*	
3	*James Smith*	18	Do	—	Do	*Alice Watts Harwich*	"	– . .	*Boy*	

Capt. Stone traded with CENTAUR to the continent. Her first passage was from Harwich to London starting 23th February 1895 arriving 5 days later on 28th February. She loaded and left London on 11th March bound for Calais which she reached on 14th March being turned round in a day to depart on 15th March arriving back in London on 18th March. She left London on 29th March for Harwich where she berthed next day. Her next passage was to Maldon on 3rd April arriving the following day turning round on 7th on a passage to Southampton where she arrived on 13th April. There was then a day passage to Portsmouth on 24th April where she evidently loaded to depart on 30th April bound for Calais where she berthed on 1st May. Unfortunately the records do not detail the freights but there then followed a series of weekly passages between London and Calais, with quick turn arounds In both ports.

CENTAUR was the first of two barges launched by Cann in 1895 , the other was KITTY and the fortunes of these two barges have had a strange parallel over their 60 years trading. Both were built for Mistley owners, both came to Colchester. KITTY followed CENTAUR to be unrigged in 1955 when the first phase of their work was done. Both later were based at Maldon as charter barges and KITTY followed CENTAUR on to the market out of the charter fleet. KITTY like her sister, also has the fine bold sheer typical of Canns yard which as Fred Cooper wrote were regarded by most bargemen to be nearest to perfection, being generally fast, sea-kindly and having good stowage space.

| From | | To | | |
Date of Departure	Port	Date of Arrival	Port	Date of Departure
25/2/95	Harwich	28/2/95	London	
1/3/95	London	14/3/95	Calais	
15/3/95	Calais	18/3/95	London	
29/3/95	London	30/3/95	Harwich	
3/4/95	Harwich	4/4/95	Maldon	
7/4/95	Maldon	13/4/95	Southampton	
24/4/95	Southampton	24/4/95	Portsmouth	
30/4/95	Portsmouth	1/5/95	Calais	
6/5/95	Calais	7/5/95	London	
9/5/95	London	10/5/95	Calais	
12/5/95	Calais	14/5/95	London	
16/5/95	London	18/5/95	Calais	
21/5/95	Calais	23/5/95	London	
25/5/95	London	27/5/95	Calais	
28/5/95	Calais	30/5/95	London	
4/6/95	London	9/6/95	Calais	
11/6/95	Calais	13/6/95	London	
19/6/95	London	21/6/95	Calais	
22/6/95	Calais	24/6/95	London	
27/6/95	London	27/6/95	Calais	
28/6/95	Calais	30/6/95	London	

Chas Stone
Master

Her original owners were the Horlocks of Mistley for whom she carried barley, malt and cattle feed between that Port and London until 1933 when she was sold to Francis and Gilders of Colchester.   She continued in general trade with them and passed with the rest of their fleet to the London Rochester Trading Company when the two firms merged in 1951.   Her gear was removed in 1955 when she became a lighter at Heybridge Basin and she was rerigged in 1964 when converted to charter work by John Fairbrother, who with two local people formed the Maldon Yacht and Barge Charter Co. Ltd.   Fairbrother, the son of a London Doctor shipped aboard SPINAWAY C as third hand in 1948 becoming mate in Ethel some three years later and in 1952 taking command of REPERTOR returning to SPINAWAY C in 1954.

KITTY stayed on the east coast until she was sold in 1977 to south coast owners for use as a restaurant at Hayling Island. On her passage out, she broke a leeboard in the Thames estuary, was towed to Ramsgate and went ashore in Chichester Harbour. In the mid 80's she was bought by Patrick Keene and moved to Ocean Village at Southampton from where she was occasionally taken out sailing. In 1990 Roger Marriott purchased her and she moved to a berth alongside the Boardwalk near the Mermaid Pub in Port Solent where she is used for corporate entertainment anniversary and similar charters. She had a new suit of sails and was again registered as a British Ship under the name of MY KITTY but Roger was able to obtain an affidavit from the skipper of the Humber barge KITTY that she had been broken up and the barge regained her original name. She is very much a family venture with her skipper John Metcalf, formerly a master mariner with Gulf Shipping and one of the family who owned the ARDWINA and the Crawley river tanker fleet. Although spending most of her working life on the east coast, KITTY went across to Cherbourg and the Normandy beaches in 1993.

Three other barges launched in 1895 survive under sail one hundred years later. ORINOCO of London, Official No. 104862, 70 ton, was built of wood at East Greenwich by Hughes and owned by the cement manufacturers Mason of Waldringfield, and later by Cranfield Bros., the millers. She was eventually sold to LaurieTester and rerigged at Hollowshore in the '70's. On an exceptional high tide she was stranded ashore but was eventually sold and refitted as a private yacht based at Hoo with Robert Deards as racing skipper. THISTLE of London, Official No. 105727, 82 ton. was built of steel at Port Glasgow by Hamilton and traded with H.Covington and L.R.T.C. with a Kelvin auxiliary installed in 1948. She was bought by Charlie McLaren and was used as a housebarge at Battersea old Church until rerigged at Greenwich during1988 and based at Shotley Marina.

VICTOR of London, Official No. 105762, 56 ton. was built of wood at Ipswich by Shrubsall. Owned by Owen Parry she was used for the carriage of cotton seed, rapeseed, ground nuts and linseed from London Docks to Colchester returning with oil in barrels for paint and varnish works. Her war service was as an ammunition store and she became a motor barge in 1947 and remained with L.R.T.C. (who had taken over Owen Parry's fleet) until 1964. She was bought in 1974 by Owen Emmerson to rerig as a charterbarge, was successfully raced and based at Upnor. The last barge built in 1895, albeit still unrigged, is the wooden OXYGEN of Rochester, Official No. 104329, 69 ton, launched at Rochester by Gill & Son. Her owners in trade included Chubb Horlock and G.F.Sully who retained her fore and mains'l until 1962/3 when she became a housebarge at Hollowshore. She was moved to Maldon for refitting in 1990 but has been left on the saltings to await the opportunity to carry out further work.

2. Trade down channel 1896 - 1933

Charles Stone continued to sail as skipper but had some changes in the crew of CENTAUR in 1898. His mate J. Alliston, a 29 year old from Mistley who was paid £3.10 per week, was discharged in London on 14th September and his replacement H.W.Hammond aged 25 was formerly on the big ketch SUNRISE of Rye. He was himself replaced in 1899 by James Dwyer aged 24 from Waterford, formerly of the GEM at £3.15.0.per week. Third Hands came and went as was their wont, W.Hagger aged 20 from Ipswich who was paid £1.10.0 p.w. left on 9th February in London to be replaced by Harry Best age 22 of Harwich from H.M.S.Rodney who made only the trip to Dover and Calais before being replaced by 20 year old Louis Walter from Germany who lasted for only one trip from 4th-22nd March. He was followed by 19 year old George Hornby from Harwich formerly of the ketch barge LLAMA who lasted until 27th April when 18 year old Robert Mullett from Mistley joined Capt.Stone from the DOROTHEA (at Lower Upnor, 1994). He lasted until 22nd August, leaving in Harwich to be replaced a week later by H.Taylor, also aged 18 from the 1866 Rochester built THE EXCHANGE of Rochester who made only one trip to Dunkirk to be replaced along with the mate on 14th September by 17 years old Harry (or Henry) Last of Saxmundham from the THISTLE of Colchester, a huge 'coal-cart' boomie owned by Smith of Burnham. He and James Dwyer stayed with Chas Stone during the following year.

Stone Brothers both traded and raced CENTAUR and she is well known as the winner of the Harwich Barge Race of 1898, when sailed by Jim Stone. The match was part of the annual Regatta and was reported in *The Harwich and Dovercourt Newsman* for 27th August, 1898. It was held off the Esplanade Harwich on the previous Wednesday and was generally admitted a great success with excellent management from start to finish. In the early part of the day the weather looked threatening but fortunately, though not much sun, the sky cleared and right through the day, fine and agreeable weather prevailed. The Esplanade was gaily decorated with flags and on the Green there were a number of round-a-bouts, swings, shooting galleries etc. which gave animation to the surroundings. The programme contained a very full list, and several events were gone through with excellent precision so that there was no waiting whatsoever. In fact at times more than one event was going on simultaneously and the thousands of people who lined the Esplanade and Green were more than pleased with the completeness of the arrangements throughout. A very amusing feature, not enumerated on the programme was the constant fun provided by five naval men dressed in gay costumes from a decorated raft. The proceedings were much enlivened by the band of the First Essex Artillery Volunteers playing from 2-6.30 and 8 - 10. The success was mainly due to the exertions of the Secretary, Alderman H.G.Everard and assistant Secretary W.Bevan supported by officers including S.R.Groom, Capt.Murray, W.Middleton, W.H.Orvis, H.Lewis and both J.Cann and H.Cann.

At 11.00 the match for barges belonging to the ports of Harwich and Ipswich was started, with the first prize 1st £20, 2nd Prize £10 and 3rd prize £5. The course was from the Committee Boat to the Cork Lightship, back to the Harbour and round a mark boat at Statton Bight in the River Orwell (twice round). Five barges entered and started, PRIMROSE (W.Wrinch) CONSUL (Groom & Son) ORINOCO (F.W.Mason) PETREL (F.W.Mason) and CENTAUR (H.Stone). Slow progress was made in the first part owing to the light wind. The wind however freshened and CENTAUR finished just before 6 o'clock when there was a good distance between all the vessels, CONSUL was second, and ORINOCO 3rd.

While the barge match was in progress there were various other events. At 12.00, midday there was the start of a match for Una Boats; at 1.45, a handicap class for yachts not exceeding 7 tons; at 2.00 a handicap sailing match between the Royal Navy, Coastguards and Revenue; at 2.10 watermen raced their Lugsail boats and at 2.20 other lugsail boats not exceeding 18ft. set off. There were rowing events as well; at 2.30 the 4 oar boats set off and so on into the afternoon when at 4.25 there was a swimming match for men. Land sports included races for men, boys and girls, a tug of war etc. etc. In the evening at 8.30 prompt there was a fireworks display on the Tower Hill, a position from which an uninterrupted view was had by the thousands of spectators assembled.

The event was a financial success and a fortnight later the paper published the accounts showing £2 contributed towards the Barge Match fund by Lipton & Co and one guinea each from many owners including W.S.McLearon, W.Wrinch, and other supporters including Farnfields (the Thames Match solicitors) and Whitmores, the sailmakers. In all £42.11.0. was contributed which with the entrance fees of five guineas totalled £45.14.0. out of which £35 was laid out in prizes and the accounts were audited and dated 6th September 1898. Some reports of the match also mention IDA owned by R.Paul, and IVERNA owned by Holmes of Harwich as participating.

In the appendix are details of the voyages for 1898/9 including that in August, 1899 when during a passage from Mistley to Hull to load wheat she lost a leeboard and her bowsprit off Sherringham. She bore up for Great Yarmouth and was towed in. In 1901 she 'sat' for a portrait by the well known American seaman - painter John Henry Mohrmann (1857- 1916) who had settled in Antwerp in 1890 and charged the equivalent of £5 for each oil painting.

On 28th April 1902 CENTAUR sailed from Shoreham for Burght in Belgium with spirits of salts under the Command of Capt.Shine. Her steering gear was totally disabled and she was towed into Newhaven by the iron steam tug BELLE OF THE USK. It took Capt Shine until 6th May to organise the repair of the barge and leave Newhaven. Her master on this occasion was probably Dennis Shine who was lost when the old ironpot FIREFLY foundered with all hands in heavy seas off St.Catherines on 9th January, 1906 with roadstone for Langstone from Cherbourg.

In 1902 Capt. Stone purchased BUCEPHALUS of Harwich, 109885, an 80 registered ton spritsail barge built that year in Belgium but she did not stay long with him, as the following year her managing owner is given as Henry J.Pulsford of Poole. Hervey Benham states in 'The Big Barges' that she proved awkwardly large to work above the bridge at Lewes to where he was then working. In October 1903 Capt. Stone disposed of his remaining shares in CENTAUR and is appointed managing owner of the new 70 ton barge, HETTY also built on the continent by A.J.Otto & Zonen at Krimpen near Rotterdam in 1903. She was still larger than CENTAUR being 90.5 x 19.85 x 6.8 with an elliptical stern and the typical bow of a dutch klipper. Other owners were William John Moore and William Hall both of Dovercourt with 21 shares each and the carpenter Thomas Sherrard of Mistley with the remaining 2 shares. The barge was sold to F.W.Horlock in 1912 who kept her until she was sold to Wynfield Shipping Co. of Grimsby shortly before she was stranded off Oronsville on 28th January, 1917.

The Stone brothers, eight of them, all sailormen like their father Charles, hailed from Mistley. They had the 200 ton (deadweight) boomies HAROLD and ELIZA H. (Jim Stone) and GENESTA (Harry Stone) and five spritties EXCELSIOR, JACHIN (later renamed VENTA), FLORENCE and MYSTERY as well as CENTAUR. In 'The Big Barges', Hervey Benham mentions the brothers, Harry, Charles and Jim. Harry was skipper of Pauls 136 ton boomie barge, IDA of Ipswich, reputedly an ugly old thing, and then skipper and part owner of GENESTA before going into spritties. He was one of the gentleman skippers, fond of stowing away his leeboards when coal-laden in North Country ports to make his craft look grander and having a flagstaff at his house at Mistley from which he always flew an ensign when he was at home, like the Royal Family.

The third brother, Jim was chiefly noted as one of the champion racing barge skippers taking CENTAUR in the 1898 Harwich race. He was skipper of ELIZA H. owned by John Holmes of Harwich which he lost on the Longsand in 1902 with a cargo of coal on passage from Goole to Saint-Valery , the crew of three rowing to the Sunk light vessel. He later lost the HAROLD which he also abandoned when she got ashore which so upset Harry (who had shares in her) that the partnership broke up as a result. Jim Stone later sailed Horlocks steel built REMINDER to victory in the Thames and Medway matches of the 1930's and as a frail old man in his 90's was seen on the Committee Boat in the Coronation Match. Frank Thompson of Layer de la Haye reports that their brother Bill who married his great Aunt had Howard's JACHIN new but sometime before 1908 retired from barging and took the Old Ship Inn at Heybridge Basin.

In the early 1900's CENTAUR began her long association with Colchester and the Colne. William Edward Rogers, known to the sailormen as 'Dolly' Rogers took shares in her along with George Langley, the licensee of the "Duke of York" Public house in Barrack Street, Colchester. Lloyds List reports that on 21st January 1905 CENTAUR was towed into Portland with a broken sprit and other damage when bound from London to Exeter with sugar. No suitable spar could be found at Weymouth and Capt. Rogers decided to accept a tow to Exeter from the tug PETREL at the cost of £15. Another report in Lloyds List on 15th December 1906 shows her 'over the other side' bound out of Antwerp when she lost both anchors off Terneuzen, Westerschelde, in the Netherlands and had to put back. The report described CENTAUR as a British Sloop. It was also in 1906 that Rogers bought a further share and became the managing owner, in this he was financed by John R.Carter a yachtmaster who provided £300 at 6%. When Carter died the loan was taken over by J.R.C.Carter of Cowes, Yachtmaster to H.M.King George V.

William Rogers continued to take CENTAUR to foreign ports carrying roadstone. On 9th December 1908 and 4th May 1909 there are records of him arriving at Heybridge Basin with 140 tons of broken granite. The following August she was working to the Phoenix Ironworks, in Lewes with 'Long Iron' from the Continent and although Capt. Stone had sold CENTAUR he was also working in this trade with HETTY. The ironworks were above the low single span arched bridge in the town and both barges appear with another sprittie in photographs of the Ironworks river front reputedly taken in August, 1909.

The photograph shows CENTAUR with her original coasting windlass and Barry Pearce points out that when it was replaced the bitt heads were sawn off level with the deck planking and new windlass bitts dropped down inside the remains of the original ones. When going down the forecastle there used to be ten inches of thick vertical oak each side. In the pictures the forecastle scuttle hatch is on the fore hold, being painted or perhaps the crew just needed more light below. KITTY had a similar scuttle but it had a round hole in the deck surrounded by an iron ring on which an iron lid fitted. When John Fairbrother renewed deck planking under the windlass the round hole was replaced with the usual square access. Barry wondered if CENTAUR had a similar round hole originally.

Centaur waiting to clear the bridge at Lewes, Sussex Photo from the collection of Hervey & Son (Lewes) Ltd, The Bridge Wharf Brewery, Cliffe High Street, Lewes.

Only one more Half Yearly Declaration held at the Public Record Office at Kew is signed by Capt.Rogers. This details the various passages made during 1911 and clearly shows the difficulty he had in earning a living in the coastal trade.   He gives the name of Clifford James Smith age 28 of Mistley as his mate at £3.15.0 per week who left the barge in London on 17th July, and was temporarily replaced by a Mr.Matthews.   Capt. Rogers sailed from London on 2nd January arriving in Dunkirk on 9th where CENTAUR unloaded and lay waiting for work until 24th when she set sail, returning light to London next day.   No movement is recorded until 13th February when she sailed for Salcombe where she eventually arrived on 7th March.   No freight was available there so she sailed on light for Plymouth arriving there 15th March.   It seemed to be another unfruitful voyage and it was not until 3rd April that she left loaded on a passage across to Belgium.   She berthed at Vilroorde some 30 miles above Antwerp on the Brussels Canal on 15th April to unload.   She remained at there until 22nd when she left without a taking on cargo, probably under tow up-stream against flood water for the Apollinaris water town of Remagen in West Germany, arriving there on 28th April.   She loaded, in all probability crates of mineral water in bottles and sailed on May 13th arriving in London five days later, probably to unload at Lucas & Spencer's wharf behind Cherry Garden Pier.   Her next passage was to return to the Brussels canal, leaving on 3rd June arriving at Burght on the 6th; sailing light, again for Remagan on 10th, where she arrived on 18th June. She loaded and Capt Rogers sailed for London on 19th June arriving on 6th July to unload CENTAUR for the last time.

Her days of regular cross channel work had come to a close and on 29th August, 1911 Capt Rogers sold CENTAUR to Edward Alfred Hibbs, sailmaker of Brightlingsea.   Ted Hibbs appointed his partner, W.E.Jarvis of Hythe Quay, Colchester as managing owner and Capt. Rogers transferred his remaining shares to finally sever his conection with his barge on 19th September, 1911.   The previous day her new manager appointed as master 54 year old Charles Caney of 25 Morant Rd. Colchester, formerly of the DIAN of Cowes and Ben Covington, born 1878 formerly of the SIRDAR of London became her mate.   Capt.Caney's grand daughter Mrs. Stebbings, who married into the family of the Heybridge Basin boatyard owners and is now aged 86 remembers when she was a girl being taken sailing by her grandfather on CENTAUR from Colchester to Wivenhoe.   The Half Yearly Declaration for the second half of 1911 signed by W.E.Jarvis gives details of the barges last continental passages but is silent about her new trade.

Bill Jarvis was in partnership with Ted Hibbs, sailmaker, chandler and barge owner of Brightlingsea and had a fleet of barges which worked for the oil mills of Owen Parry and for a time CENTAUR joined his fleet.   She worked along with GRETA and MILLIE which were built for the partnership, the BURNHAM and the little UNITY and 1915, the date of the next Half Yearly Declaration held at Kew, she made regular weekly passages from The Hythe Colchester to Millwall. Sometimes sailing light drawing only three foot or with a part freight she returned loaded down to a six inch freeboard. The fleet brought linseed and cottonseed from London and took back linseed oil in drums carried below deck. Towards the beginning of the First World War, freights to the oil mill were falling off and after the war started the highly profitable trade in coke and pitch to Le Treport, Calais, Dunkirk and other continental ports was built up. The Brightlingsea coasters like ALARIC, LESLIE and MAJOR joined CENTAUR and many barges in this trade and little wonder, 125 ton of coke to Calais or Boulogne at nearly £2 a ton was being paid in 1918!

CENTAUR was always very well turned out at this time. Felix Mallett recalled in 1975 when aged 93 that there were no barges as well maintained and smart as the oil mill barges, none that is except for CENTAUR and she was one of the clippers of the Colne. Indeed, according to Rod Minter, Ted Hibbs had bought CENTAUR from Dolly Rogers for £550, and in just over four years, sold her to John Sawyer the Brightlingsea barge owner and licensee of the 'Yachtsmans Arms" for £800, she was so highly thought of and valuable a barge.   The records at the Public Record Office actually show that in 1911 Hibbs transferred 21 shares to William Stokes Jarvis. a further 21 to various people and in 1916 Owen Parry Ltd. became the owner of all 64 shares. After nearly two years they transferred the barge to John Sawyer, described as a Brightlingsea Fish Merchant but within a month he had transferred three quarter shares to become only an equal quarter owner with William Pattison, Arthur Butlin and Edward Pudney but the Sawyers, father and son had her for the best part of the next ten years.

She was now working out of Colchester and as River Dues were paid both on inward and outward passages there was much to be gained by keeping down the net registered tonnage particularly as dues were at 2d per ton for craft between 50 - 59 g.r.t. whereas they were payable at the 3d per ton rate for craft between 60 - 69 g.r.t.   Sawyer had some minor alterations made.   The barge was resurveyed on 25th March 1919 and he tonnage was now below 60 registered ton!   The original registration certificate issued to John Sawyer has now been folded in two and attached with a pin to the file of papers kept at the Public Record Office at Kew.

Just before the War the well known Skipper Captain Ephraim "Chick" Cripps of Foulness began an association with CENTAUR that lasted for over twenty years. Chick Cripps came to Colchester in the barges of Ambrose Ellis of Stanford-le-Hope who traded to all parts of Essex.   His first command was the MATILDA AND AMY, later he had the MINNIE and then ALASKA.   He took the CENTAUR for Hibbs for a while but left her for a spell of about three years coasting and running across the Channel in the KLONDYKE, which was as much at home in the Humber or the Schelde as in the Thames.   He became based in Colchester, being skipper, first of the LESLIE WEST and then, again and finally in 1922, of CENTAUR of which he thought so much.   During his first spell as skipper, in the well known great gale of 29th December 1914, CENTAUR lay in the Colne laden with cottonseed for the oil mill.   The Maldon barge ROSE lay nearby wheat laden. During the gale both dragged rapidly in towards the beach.   Several times CENTAUR and ROSE crossed each others paths, narrowly avoiding collision, the skill of Chick Cripps on CENTAUR and Sam Wallace of ROSE averting a disaster which would have inevitably sunk one or both craft.   Oddly enough our barges namesake CENTAUR of Rochester also lay in Colne that night. Not to be confused with the Club barge, this one was well known for being built in six weeks before the 1899 races.   She broached broadside on and the first wave to hit her cleared her deck of her cargo of drums which went floating away into the Colne.

The half yearly Official Log Books which Ephraim Cripps completed for 1915, 1925, 1928 and 1930 giving details of voyages which survive in the Public Record Office at Kew show that he was born in 1882 and that he had a home address at 2 Ripple Road, Dagenham.   The barge's owner at the time of the June 1915 return was shown as Edward Alfred Hibbs of Regents Road, Brightlingsea while in the following December the owners were stated to be Messrs. Jarvis & Hibbs of the Hythe, Colchester while John Sawyer of 10 Colne Road, Brightlingsea is shown as the Managing Owner in the later returns.   Chick's mate at the time of the 1915 return was Charles Good of Beaumont, Essex who was only two years his junior while in 1925 and 1928, Harry Piper of 'Kent', born 1904, is named as mate. Throughout this period records show the barge engaged in trade on the east coast.

Some of Chicks exploits were recounted by Mr.L.A.Piper recalling that his father Harry Piper, who was as born in Kent in 1904, ran away to sea, or to be more precise went to Rochester.   He joined CENTAUR in 1919 or 20 after his father insisted that he refuse an apprenticeship.   During his first voyage to Boulogne they ran into bad weather and he became violently seasick.   Despite this discouraging start he remained on CENTAUR for a number of years until he became skipper of the SALTCOTE BELLE, swallowing the anchor in about 1950 when her owners proposed fitting an engine.   In August, 1921 "Chick" and "Toby" Piper were in the Tyne when the airship R38, the largest airship then built, embarked upon a test flight prior to crossing the Atlantic for her new American owners.   She left her base in Bedfordshire on 23rd August, 1921 and proceeded up the east coast of England recording a commendable 71 m.p.h., until by early afternoon on she was over Newcastle.   At this point she executed a series of violent turns, why was never established, but she broke in half and exploded.   Pieces of burning wreckage rained down and one large large piece of the tail brought down with it two senior members of her crew. The broken, twisted wreckage landed on a sandbank in the lower Tyne estuary, fortunately at low water.   Both men were injured, one unconscious and as quoted in a radio programme "Scrapbook for 1921" were lucky to be alive although they were still in great danger, entangled amongst the remains of the airships wire rigging, unable to move. "The tide was now rising when, just as it seemed certain that we should both drown two unknown men in a rowing boat appeared from nowhere and rescued us". Chick and Toby freed the injured men and having taken them to hospital and not wanting a lot of fuss and anxious not to get involved with official enquiries they quietly slipped away. Of the 49 men on board the airship only five survived.

Mr.Piper also tells a different story involving Sir Alan Herbert who had a passion for any kind of water craft and in the twenties often visited sailing barges on the Thames, sometimes holding impromptu parties on board with his theatrical friends. Occasionally he went for trips with them and bargemen were always pleased to have him around as he was good company. He was also an extremely useful man to have on board; he was very good with bureaucrats having been trained as a barrister he tied them in legal knots. On one occasion he sailed on CENTAUR from London to Brightlingsea, and on reaching the Colne estuary they dropped anchor. Before very long a police boat appeared and came alongside, an officer climbed aboard breathing fire demanding to know why they had anchored over the oyster beds accusing them of stealing oysters. At this point Alan Herbert emerged from the cabin asking what the trouble was. Ten minutes later the policeman departed unable to get away fast enough, apologising profusely for having disturbed them, his head in a whirl from legal arguments as to the ownership of oysters, the laws of navigation, the penalties for defamation of character. etc. etc. It was perhaps just as well that he'd failed to notice a rope disappearing over the side of the barge on the other end, out of sight, dangled a galvanised bucket containing a dozen oysters destined for Alan Herbert's dinner.

While Harry Piper was still serving as mate in 1928 when Chick completed another of the pairs of Official Log Books now held at the Public Record Office, Charles Sheldrick (born 1895) formerly mate on SIRDAR joined Chick at Colchester on 1st January 1930 but as detailed in Appendix 4 the work remained very much as before. Chick Cripps was a nephew of the owner, John Sawyer who in the 1920's gathered several barges together for use in the general coasting trade with the help of various partners. Indeed in 1923 Chick Cripps became part owner of CENTAUR taking a quarter share in her and to a large extent using her as his home as he remained a bachelor.

Around this time another one quarter share was bought by the partnership of Joshua Francis and Cecil Gilders who were putting together their own fleet and this was the start of the next chapter of the story of our barge.

3. The days of Francis & Gilders 1933 - 1955

In 1915 Joshua Francis returned to Colchester to manage the fleet of 6 barges, owned by his uncle, Henry Howe. These craft, KEEBLE, FEDERATION, SUPRISE. PEACE, BRITISH EMPIRE and FALCONET were distinguished by their black topsails. 'Josh' was the son of a barge skipper and had gone to sea as a boy in barges and tugs. Another uncle, David was also sailing out of Colchester in the little steam hoy GEM but according to Hervey Benham writing in the T.B.S.C. Journal, *Spritsail* in the Summer of 1949, Josh shipped in a Kentish barge and settled for a time at Aylesford. In Kent he took the SUSANNAH for Silas Wagon and the KENT and BESSIE TAYLOR for Store & Co working between London and the Medway. Once Josh completed 14 freights in seven weeks, a feat only possible because loading and unloading were done at the same wharf at each end. Then he turned over to tugs, first for Store & Co and then taking a "Kaiser" (a title which had to be dropped abruptly in 1914), for J.P.Knight.

After the death of Henry Howe, Josh went into partnership with Cecil Gilders to manage the six craft. The partnership gathered together most of the Colne and Blackwater barges, (the exception being the mill barges owned by Marriages and Green Brothers of Maldon and Brantham). From Sadds came OAK and EMMA; from Keeble came MIROSA (then known as READY); from John Sawyer came ALARIC, DELCE, LADY HELEN, LESLIE WEST and CENTAUR; Colonel Bingham brought in his auxiliaries CASTANET and BANKSIDE (both later victims of enemy action). From Eldred Watkins came the ETHEL ADA of Ipswich and GENERAL JACKSON, COLONIA, VARUNA, SEA SPRAY, CLARA, AGNES AND CONSTANCE; from Stanford le Hope, GEORGE SMEED, SALTCOTE BELLE, BEAUMONT BELLE came from the upper Blackwater and Walton backwaters, until the company owned or managed over thirty craft before the outbreak of the 1939 - 45 War. In addition to the craft flying the purple and gold bob, the company managed barges owned by Prior of Burnham and Shrubsall and by 1949 when Hervey Benham was writing only ALARIC and VARUNA had auxiliary engines , soon followed by LESLIE WEST, all with Kelvin 66's.

CENTAUR officially joined the fleet on 2nd October, 1933. Her bill of sale records her value as £800 at the time. This compares with £300 for the ROSE and £900 for

VARUNA who ended her days broken up on the beach at Worthing, her compass having gone into CENTAUR some time before. When CENTAUR came into the fleet she was primarily engaged in the East Anglian grain trade, trading to Ipswich, Yarmouth and Felixstowe, and was always reckoned to be a fast barge. On one occasion she was lying at Cranfields Mill, Ipswich along with GERTRUDE MAY and four others of Francis and Gilders' fleet. The wind was very strong and blowing straight up the dock. One at a time Hedley Farrington took the barges down to the gates. The skippers of the other craft exclaimed that CENTAUR made two trips less across the dock than any of the others.

When Francis & Gilders Ltd. was formed as a limited company on 2nd October, 1933 Chick Cripps was Commodore-skipper of their fleet and a major shareholder with 2000 shares. His mate for the last few months of 1934 was Harry Dadds of Colchester aged 28 and he stayed with Chick until his last voyage as Master of CENTAUR leaving London drawing only five foot at 6.30 on 24th April, 1935 to berth at Colchester at 2.30 on 27th April. However, it was not until 3rd June, 1935 that they were both discharged with "VG" reports for both ability and conduct.

When he retired after commanding CENTAUR for 23 years, Chick went to live in a converted lifeboat at the Colchester bargeyard and looked after the Francis and Gilders fleet as ships husband. It was largely due to his devotion that the Colchester sailing barges enjoyed a reputation for smartness and good maintenance, and with his rich store of memories and friendly personality he was also a friend of many yachtsmen. He died at St.Bartholomew's Hospital in London and was buried at Foulness. A short obituary appeared in the *Essex County Standard* 18th March, 1949 and in the Barge Club Journal the following summer.

Arthur (Sorbo) Keeble was next to have CENTAUR. He was a Maldon man, born 1884 and living at 174 High Street. He joined Josh Francis' fleet in the 1920's having previously been in PALL MALL and SALTCOTE BELLE and had been well known as skipper of the stack barge PRIDE OF ESSEX when she caught fire at Vauxhall in 1922 and burnt out with two other barges and a lighter. He then took the Murrell's barge CHARLES & ISABELLA of London but lost her on her first voyage at the Spitway. The steering gear broke, the rudder probably touched sand, which wouldn't have happened with a tiller steered barge such as the PRIDE OF ESSEX! Immediately before taking over CENTAUR he had been in AGNES & CONSTANCE with his mate, the 23 year old George Cracknell from Colchester and they took their new barge away light at 10.30 on 5th June, 1935 bound for Maldon where they berthed at 2pm that day. Arthur Keeble had the job of completing the Offical Log-Book and List of the Crew for 1935 but apart from two passages London - Fellxstowe in August and London - Ipswich in November, all other trading was in the Thames estuary.

Arthur said of CENTAUR : "Six hundred quarters is enough for her; six-fifty is too much. She has always been a fast barge, knocked 'em all a'cock at Harwich one time. She belonged to Stones of Brightlingsea and beat the ORINOCO,IDA, IVERNA, and PRIMROSE in the Harwich Regatta. She's strong, why all the floor timbers in the middle of her are a foot square. There's strength in her dead wood too. That comes right aft into the forehold, and t'other end, that's scarfed in under the cabin stove."

As mate on CENTAUR, Arthur Keeble had Jimmy Polly from Ipswich who became very much a star turn among the Colchester skippers and mates. Various tricks were played on the unsuspecting skipper including dropping kippers down the chimney as a result of which she reeked of fish for days afterwards

Arthur Keeble on Pride of Essex

At the time of Dunkirk in 1940 CENTAUR was towed down to Dover with the LADY ROSEBERY and DUCHESS to take part in the evacuation of the British Expeditionary Force.

Vice President Arthur Bennett describes her part in the operation in an article about the *"Maldon Barge Master"* in the Club Journal for the Winter 1965-66. Arthur Keeble told our Vice President how she was lying in the Prince of Wales Dock alongside the yacht barge JAMES PIPER who had been on her way round to the Thames from the south coast, when a tug came in at an excessive speed.

> "Some of our chaps started a-singing out. 'Don't you worry, I said. 'He won't hit us. He can stop her.' But he didn't! He came into us head first.
> JAMES PIPER being on the outside, got her coamings sliced through.

But for a good ol' bumping, it didn't seem we'd taken any harm, though we had a good look round.   Presently soldiers came aboard'.

'We're going to start a-loading on you now, Skipper.'

'That's all right' I says, for we'd already uncovered. Presently though, one of the soldiers come down aft   'There's water a-coming into your barge' he says. "I knew that, for they'd started making a bulkhead of cans of drinking water under the mastcase.

"No, no' he says, 'She's leaking, skipper.'

"Dear O dear! So she was, too. Water over the ceiling in the hold, and we had a rare ol' job a pumping. Me and my poor ol' mate were at it all night long. A naval officer came along and took a look at us. The soldiers worn't to put no more aboard, he said. That stuff was too valuable to chance losing it, so they loaded it into the LARK instead."

The jerrycans reached Dunkirk but little, if any, was drunk; LARK was abandoned on the beaches and discovered by a group of soldiers and sailed back towards England. When the navy took off the survivors LARK was sunk by gunfire so it is perhaps as well that CENTAUR was unloaded in Dover!   Back in Dover, it seems that CENTAUR's chine had been damaged against the sloping dock wall but there was nothing they could do there.   Arthur Keeble said they'd better go and get repairs done.   They wanted to route us to Gravesend.

"That's no good to us' I said

"Where d'you want to go then?"

"Why, back to Maldon. We've got a yard that'll see to us there'

"All right, they said, 'Clear off"

CENTAUR eventually sailed for Maldon at about ten o'clock in the morning, with an extra hand on board to help pump (according to Frank Thompson this was Russell Dent, skipper of the Maldon EMMA who went with them backs to Cooks Yard at Maldon). They shaped up for the West Last then stood on for what they took to be the Mouse and the Blacktail.

"That come on thick as guts. Worn't Mouse! Worn't Blacktail. We
fetched `up alongside the Sheers. Tide was a-flowing by this time and
we came tight over the top into the Blackwater and brought up under
Wymarks between eleven and twelve that night. Never saw a thing!"
(Wymarks is the land below Bradwell, named after a Saxon given the
 land to farm, the anchorage off the shore has had that name since then)

Apart from this incident Arthur Keeble had CENTAUR in the East Anglian work during the war.   Brick rubble from the blitzed areas of London to Maldon for the construction of the aerodrome gave employment to a number of barges; there was a great deal of laying about and some of them made more on demurrage than they did from freight.

"Same as at Ipswich. We loaded up at Millwall for Ipswich along with DEFENDER, OXYGEN and a whole lot others. They all went straight in and tied up at the abutment. When they called out to them to come alongside and load they took no notice, so they loaded CENTAUR first".

`"A wonderful lot come away from town and brought up under Shotley, so we did the same. George Bowles was in VERAVIA and wouldn't muster 'till we were a mile past. Up at Ipswich I said they'd come first open turn, but they wouldn't have it. That was a proper racket, to leave barges on demurrage. I told the foreman I'd report it to the Ministry of Transport, but he said there was no need to carry on so".

"Why should I pay you five bob to take other barges out of turn?" I said, That's not right, and nor it wasn't, not that it did George Bowles much good that time for they got him out before the week-end anyway, and he lost a freight, what's more.

Frank Thompson recalls meeting up after his demob with Arthur Keeble and hearing of the death of Jimmy Polly during the war, Sorbo then had as mate, Frankie Dowsett, previously mate of the SEA SPRAY. Sorbo shifted from CENTAUR in 1948 into the 120 ton FALCONET, built by Curel in 1899 as she was easier to work. He retired in June 1950 at the age of 65 to the same house in Maldon High Street that Strutt built about the time he ordered the SURPRISE, the first barge built by John Howard, in 1878. When the cottages were pulled down Barry Pearce aquired the front door of No.174, still with the knocker that Arthur used to rattle all those years ago.

CENTAUR was largely in the grain work, mostly between London, from the Royals or the East & West India Docks and the East Anglian ports of Colchester, Felixstowe and Ipswich with an occasional freight down the coast for Great Yarmouth. When the Maldon work fell off there was wheat for one or other of the mills from time to time, and timber for Sadd's, either a trans-shipment job from the Surrey Commercial Dock or lightering from some Scandinavian brought up off Osea. Francis & Gilders were left as the last 'seeking' fleet, that is their craft carried any cargo they were offered, anywhere on the east coast. The masters of their barges could fix their own cargoes if the opportunity arose but the Company arranged freights through agents in their Dominion House office in the city. The agent passed on the freight fee to the bargeowners after deducting his commission and the owners deducted any towage charge, dock and light dues and their share before accounting to the master who passed on one third to the mate keeping his two thirds. Once a year the barges went on the blocks, usually at the Company's yard above the bridge in Colchester, or at Cooks yard at Maldon, the gear lowered down, the sails taken ashore to be checked over, and the crew paid £6 a week.

Captain Hewson took over after 'Sorbo' Keeble. In "*Us Bargemen*" Arthur Bennett writes of meeting Capt. 'Ewson in Whitstable at the end of August, 1948 and says "he had a voice like a fog horn and suffered no inhibitions". He came from Northfleet, had been a lighterman and freeman of the River and had been in barges belonging to Wills and Packham and Everards. The second edition of *Spritsail* notes that Capt.Hewson had just sailed CENTAUR light from Chatham to London and then on the Whitstable to load Kentish wheat. Before that he had the TEETOTALLER for a time working to the Queenborough Glue Works and blowing out of Sheerness. One day with barrels in the hatchway, he was accosted by a yachtsman who yelled after him, "You're a fine one; TEETOTALLER and got a freight of beer and all!"

Then followed Captain Michael Blythe who had several of Josh Francis' barges, many for short periods, and was one of the earliest skippers to use an engine: a motor cycle engine fited to the leeboard of ALASKA. He had CENTAUR just before Stan Yeates took her over in 1950. Stan had as mate Bill Furze previously mate of the JOY followed by another young lad, Billy although Stan always sailed with his wife, Chick who kept diarys of their days in GENERAL JACKSON, SALCOTE BELLE and CENTAUR.

In "*Us Bargemen*" Arthur Bennett writes of meeting Stan, his wife, Chick and Jack their then mate in early August, 1950 at Maldon when HENRY, festooned with bunting for the Maldon Town Regatta, lay outside the Jolly Sailor. CENTAUR had sprang a leak in the Wallet and had returned to Maldon to have her bottom re-fastened at Cook's Yard. There were plenty of visitors at the Maldon Town Regatta including John Glenn who had been crewing his friend, Bill Haines, a fellow member of Greenwich Yacht Club in his converted lifeboat MORNING MIST. Bill Haines was a friend of Arthur Bennett and both he and John were invited aboard HENRY where they met Stan and Chick. The next day CENTAUR was leaving to act as the committee boat for the regatta and Bill and John begged a passage along with the Bennett family and Arthur's daughter Elizabeth. Anchored off Mersea, Bill and John took Chick ashore in the boat to arrange the committee boat mooring off the Mersea Yacht Club. They had a heavy pull to get back making about a quarter of a knot over the tide. The next morning she was towed in and decorated over all. They had a grandstand view of the 'greasy pole', 'miller and sweep' and other events ending up with fireworks. Stan and Chick had a very

Colonia and Centaur loaded

enjoyable weekend.   Stan and Chick were very kind and invited John to join them on their next trip which was from Fingringhoe Mill on the Colne with wheat for London.

Sometime later Frank Thompson, who also had several trips with them, gave Stan a photograph which he had obtained from the Essex County Newspapers showing Hervey Benham standing next to the skipper with about a hundred people on board (much later a copy was on show in her saloon).

Frank ruefully recalls that he was foolish enough to call her a sluggish old cow when turning her up through Greenwich Reach and as a result she broke his finger at 3am when locking into the Greenland Dock! Stan and Chick made about £35 for each freight in times varying about a fortnight, which was not bad money at the time and Chick was soon acting as mate.

John Glenn writes from his home in Grantham "I had been a member of Greenwich Yacht Club and active on the river since 1933 and had sailed before on a barge on the London garbage run to Mucking Creek, so I was probably able to help a bit; thereafter they welcomed me aboard whenever they were in the docks at London. My son Stephen's earliest memories include being taken to the docks under the grain hoist. I was a teacher at the time and was able to do the occasional passage with them at half term and so on, helping out as required.

Unloading fertilizer Ipswich Quayside, 1950

A confession that I am prepared to make now after forty years is that one appeal of CENTAUR (apart, of course, from the charm of the skipper's wife) was that she was the only barge I knew on which I could work efficiently as a mate, because she had in place of the usual brail winch, a big affair with barrels and pawls all over it that took the topsail up on a wire halyard. Without it, I should never have had enough in me to get the topsail up taut! She had a splendidly roomy forecastle with a coal fired range, but, of course, accommodation was not all that spacious even for Stan and Chick. I linked up with the Thames Barge Sailing Club in 1951 when they had the ARROW on the river frontage of the pub which is now much tarted up and called the Cutty Sark, and joined Stan again for six weeks that year when he was left without a mate. My first job was to clean up the forecastle but thereafter it was all good bargework, in a life that was crumbling about our ears"

Stan himself writes "As for the old CENTAUR she was a happy ship and apart from one incident nothing much happened while my wife and I were in her. We had some good sailing in her and managed to average about one freight a fortnight throughout the year. The aforementioned incident happened in the January of 1952. We ran down Swin with a nice westerly wind bound for Felixstowe with 600 quarters of wheat from the Surrey Docks, and were just shaping up for the Spitway with the wind increasing a little when the short steering arm parted. We hove the barge on to the port tack and made some sort of repair with wood and wire and had another go. Halfway across the Spitway the lash-up came adrift and we finished up on the Buxey side of the Spitway. Before the flood tide came, the wind increased to about force 6 or 7 and before we picked up we had sea breaking right across the hatches. We urgently needed a tow of some sort so we put up a few rockets. It was incidentally the first call the new Clacton lifeboat had received since she came on station, having been on show at the South Bank Exhibition. She could not get away for some reason or other and so passed the job on to Walton. However, before the Walton boat arrived the auxiliary barge SAXON of Wests fleet arrived and stood by until we managed to float a line over. She eventually towed us up into the Colne. The Walton lifeboat eventually caught CENTAUR and SAXON at the Bar Buoy. We had by that time rigged up some sort of steering by using the kicking strap and mizzen sheet with a tackle on each leading to the leeboard winches".

John Glenn finishes his letter "The following year, 1953 I think it was, Francis and Gilders began to run down and Stan left them to skipper GLENMORE, a barge with a motor installed. I visited her up river at Fulham where she was working with builders' material but I lost touch with Stan and Chick as I was getting married again and leaving Greenwich. I did see them once again, when they had taken on a sand coaster engaged in a regular run between Upnor on the Medway and Rowhedge".

One of the last skippers to have CENTAUR in trade was Captain Frederick MacVicar Wilson, who along with Ebbie Shrubsall of Sittingbourne, who sailed KITTY, were the last star turns among the crews of the Colchester fleet of sailormen. Captain Wilson, known as "Nelson" Wilson was born in Folkestone on 30th October 1909 and had lost an eye playing with scissors at the age of three. He became master of the sailing barge PEARL at the age of twenty two. SUNRISE, GLEANER, ALDERMAN, KNOWLES, QUARRY and DECIMA were some of the craft on which he learnt his trade. Nelson was in KNOWLES off the Norfolk coast during the tragic storm in 1934 when the SEPOY and the HIBERNIA were wrecked at Cromer but he managed to weather the storm and sailed into a safe anchorage. On one occasion he was bound to Wisbech in KNOWLES when the trip took much longer than expected, although he was spurred on by an engine all of thirty horse power! When at Wisbech he phoned Rochester, "Please send another engine, I'm not coming back until you do".

While master of DECIMA in 1940 Nelson had his barge commandeered by the Government. He joined the steel motor barge CORSAND and went to Dunkirk and served throughout the war ferrying supplied to the forts in the Thames estuary. After the war Nelson was master of Marriages motor barge THE MILLER and then served in several barges R & W Paul's auxiliary JOCK and the big Samuel West motor barge OLIVE MAY which ran coal from Keadby to Harwich.

Capt. Fred Wilson (Nelson) London to Ipswich with maize 1954. Note the chaff - cutter wheel

While master of the JOCK he rescued some sea scouts from Erith who had got into difficulties with their boat in the Thames. Francis & Gilders obtained freights from the London and Rochester Trading Company including wheat, maize, timber and sugar. CENTAUR worked most ports in the Thames estuary and sailed as a staysail barge. She had a boltsprit but it was stowed on deck, not rigged and sailed her best in a strong breeze. Nelson found her to be a steady sailer that liked to be sailed in a certain way.

Crew at dinner in the cabin,

She seemed to prefer sailing on the mainsheet with minimum tension on the sprit vangs. She once sailed into Harwich Harbour in company with the VENTURE bound for Ipswich. Both barges had to turn up the river Orwell and using his new found formula for sailing CENTAUR, Nelson was greatly pleased to arrive at Cliff Quay three quarters of an hour before VENTURE. He was especially pleased with CENTAUR as VENTURE was always a smart fast barge. During Nelson's time on CENTAUR he met and became great friends with film maker, Nicholas Hardinge.

He recalls CENTAUR featuring in a film produced by Nicholas called "Where the Wind Blows" when the cameraman from Seven Seas Films tried to use his clockwork camera on the plunging deck. The mate of CENTAUR at that time was Mick Gibbon. Nelson was appointed master of VERAVIA in late 1954 by Nicholas Hardinge and stayed with her until she foundered down channel off Cape d'Antifer in October 1960 when reduced to a motor barge with a Kelvin 88hp semi diesel engine.

Brian Everett filming for Seven Seas Films,

41

Nick bought the motorship HERB ex Rhone and Nelson took charge of her. Apart from racing barges, including WESTMORELAND that concluded Nelsons working days but one occasion saw him sailing CENTAUR in one of the post war matches where all the crew were turned out in top hats. Unfortunately she came in last.

In March 1951 Francis & Gilders Ltd. merged with the London & Rochester Trading Company although the barges continued to wear their purple and gold bob. Her passages were now noted in *Sea Breezes*, she brought several timber freights to Colchester and Maldon and in October 1954 she sailed light to Ipswich to pick up a freight of East Anglian beet sugar which had been in store and sailed for Tate and Lyle at Silvertown. In March 1955 she made one of her last visits to the Orwell with a freight to Felixstowe Dock returning to the Thames from Ipswich. At the time she was in the ballast trade the inside of her hatch coamings, and those of KITTY, were sheathed in steel to stop them being worn away by the grab cranes when unloading. The steel sheets were not removed until 1991.

The last skipper to command CENTAUR in trade was a young Welshman, Brynley Seton Weightman whose name was given to the Registrar of Shipping on 10th January, 1955. Bryn was appointed skipper with Sammy Langford, an ex Arethusa lad as mate and joined his first command on the buoys at Erith just before Christmas. He sailed her light to the Medway to pick up further mooring warps and other gear from L.R.T.C. then up to Halling to load cement . Bryn believes that he was the last skipper to load cement under sail; a long time after Goldsmiths CARINA. After the Christmas break he carried the freight to London. He loaded 40 gallon oil drums from Grain to London and sugar from Ipswich to London and believes he took one, if not the last, load of timber to Maldon for Sadds during the three months he was her master. At the age of 17 he had been mate in GEORGE AND ELIZA and later was in ADRIATIC with Tich Irving. He was also mate for two years in the yacht barge SANTILLE ex Beryl which was converted by the Whitewall Barge Co with twin Chrysler engines and a 'conservatory on her deck. This mean that the main horse had to be set back and the only time Bryn recalls her being sailed was when his skipper was sent off to hospital in Ostend and he was able to set sail himself. When he left CENTAUR he went to Lapthornes and took the NELLIE after Bob Childs left her. He left barges to work laying carpets and to run craft fairs but retained an interest in sailing craft owning craft in Faversham and recently showed his welsh origins by demonstrating paddling a coracle!

London & Rochester did not really want to keep the old sailormen they had acquired with the Francis & Gilders company name. John Kemp who was intent on setting up The Thames Sailing Barge Trust, writes how Mr. Gill offered him any one of the four he still had at work, at £250, all standing. Kemp wanted a barge capable of carrying at least 700 quarters of wheat so he did not take up the offer.

It was a sad day for Colchester Hythe when in June 1955 the press announced that the four remaining Colchester sailormen were to go to Brown and Son of Chelmsford to be unrigged for use as lighters. MIROSA lay by the bridge, her sails blown out on her last voyage, occupying the berth where CENTAUR had several refits. CENTAUR herself was too large to pass above the Hythe Bridge to Josh Francis old barge yard which was to close.

West India Dock, loading maize for Colchester

The quartet, MIROSA, GEORGE SMEED, KITTY and CENTAUR were unrigged, the sails going by handcart to the bargeyard where Hedley Farrington recalls burning fifteen sails in a fortnight. Fittings and ironwork went to the local scrap yard. Browns acquired the four barges for £1,100 the lot. The Merchant Shipping Register for CENTAUR was endorsed with a note "Register closed this 12th day of October 1955. Vessel converted into a lighter. Registry no longer required."

As part of the takeover of his company Joshua Francis of Colchester, formerly the managing director of Francis and Gilders Ltd. became a director of London & Rochester Trading Co. Ltd.

In December, 1956 the papers recorded his death and the Company itself was eventually dissolved on 3rd December 1976.

Centaur leaves Fingringhoe jetty with Ballast for London, 1954

Upper picture shows CENTAUR at Heybridge Basin with IDA and the eel barge MARK on the opposite bank, the lower picture shows KITTY loaded with timber. Both photographs taken on 1st September 1960

4. Return to Sail.
1965 - 1974

For the next nine years CENTAUR worked as a timber lighter for Brown & Son. After a lifetime at sea she was now reduced to working between ships moored in deep water off Osea and Heybridge Basin where the loose timber was unloaded into shallow draft canal lighters and carried to the yard at Chelmsford. The timber merchants had quite a large fleet of former sailing craft including Horlocks EDME, EMILY, GEORGE SMEED, IDA, KITTY, MIROSA, MISTLEY, ROSE, and WILLIAM CLEVERLY. They were eventually joined in Heybridge Basin by the remaining Francis & Gilders motor barges, BRITISH EMPIRE, DAWN, LADY HELEN, VARUNA, and LESLIE WEST, and in 1961 by Daniels' KATHLEEN until alterations to the lock, a change in transporting and marketing of timber led to all being laid up.

Even before Browns barges were officially up for sale the former Pauls barges, IDA and EMILY were bought by three partners in Maldon; £50 per barge was offered and accepted. The remaining fleet, some still timber laden, came up for sale and some went for as little as £12.10.0. (£12.50p)! MIROSA went for £250 and LADY HELEN was sold out of the Basin for £450 to D.Hofford of Duxford, Cambridge on 26th February, 1965. The Bill of Sale describes her as a dumb barge but her last skipper in trade, as a motor barge with a 44 h.p. Kelvin was Wilfred (Willie) Charles Hardwick. She was to become a restaurant on the River Cam but for one reason or another that didn't happen and a piece of her transom with a couple of carved letters of her name sits alongside the television set in Barry Pearce's home while her bow timbers lie in the mud near Heybridge Basin. The remains of the stern and rudder rest with other old barge wreckage down at the dump beyond the sailing club compound in the saltings with the remains of PRETORIA, MAMGU, BRITISH LION, WILLIAM CLEVERLY and VICUNIA.

Barry has written that he liked the look of KITTY but noted that CENTAUR had a bad place in her chine where Dilberry Clark, who was responsible for maintaining the barge hulls, had caulked her with a lump of clay. John Fairbrother bought KITTY for £300 and John Prime the owner of GIPPING bought the second hand sailing gear to rig out other craft in partnership with Barry Pearce. One day, while they were down the Basin looking at KITTY, still piled high with timber even though sold, Richard Duke turned up as he had heard the 'lighters' were for sale. They told him a yarn that a gentleman from London was thinking of turning CENTAUR into a restaurant on the upper Thames. Capt. Duke was so upset at the thought that he bought the old barge straight away!

Richard Duke was born into a family farming in Cambridgeshire and Essex but in 1928, at the age of 13 he joined HMS Worcester at Greenhithe and went to sea as an apprentice with Houlder Brothers and later went into Shell tankers. The war saw him in Atlantic convoys bringing vital supplies to Britain under threat from U-boats and was on one occasion torpedoed but with his crew escaped in lifeboats. He was given command of EMPIRE PROTECTOR in 1942 and came ashore from the fleet auxiliaries to run the family farm near Saffron Walden in 1947. However he retained an interest in the sea and when the I.C.I. fleet came up for sale he bought the ETHEL ADA of London, which he quickly sold on, and MILLIE which he sailed both as a yacht and as a charter barge. In 1966 he bought the LESLIE WEST with the idea of restoring her to sail at Pin Mill but she remained a comfortable second home on the foreshore under the trees at Pin Mill.

Richard Duke at the wheel of CONVOY.

He planned to use CENTAUR in the same trade and she was towed to Pin Mill where most of the sailing gear out of David Antill's MAID OF CONNAUGHT was put into her. It was a speedy refit with her main mast originally coming from THALATTA and her main sprit from PORTLIGHT when she was turned into a fully powered motorbarge. The sprit had a kink in it about 10-12 foot up where it hit the rail on its way overboard when the steel barge was derigged. During 1965 and 1966 CENTAUR was owned by jointly by Duke and Antill and was skipped by Mick Lungley, although the Blackwater and Medway programmes show Derek Ling as her skipper with Mick Lungley taking MILLIE. John and June Prime acted as booking agents from their homes in North Street Maldon dealing with charters for both KITTY and CENTAUR.   John and June had bought GIPPING out of trade from I.C.I. and kept her for seven or eight years keeping in touch with Ron Goodwin her former mate, the son of "Stratford Jack".   They used their barge as a sailing home before coming ashore and organising the chartering kept them in touch with the barge world.   Although Richard Duke had put up the money to buy CENTAUR, he was tied up with MILLIE and later with CONVOY so her conversion to a charter barge and the day to day management was left very much to John and June and they all met up only for the formal meeting of shareholders and at race times.   John was also involved with John Fairbrother and Barry Pearce with the rerigging of KITTY and the internal layout of both barges was very similar although CENTAUR had a larger engine room which doubled as a workroom.

Eddie Smith who now lives at Shotley writes "I moved to Maldon from London in 1966 and John Prime asked me if I would go as mate the next season and help to fit the barge out. Tommy Baker sailed the CENTAUR round from Pin Mill to Maldon that Autumn together with John Prime. That weekend I went aboard for the first time and remember Tommy telling me how little wind there had been on the trip round. They had stayed overnight at Stone having been beaten by the tide and got underway from Stone at low water and had still nevertheless taken the entire tide to get to Maldon. The engine was installed by Tommy Baker himself and I believe that it was John Pitt who bored for the stern tube.

At that time the barge had a more or less open hold as from cargo days, the only change being the erection of the small toilet cubicle forward on the starboard side and a doorway which had been cut into the fo'c'stle bulkhead to allow easy access between the fo'c'stle and the hold. Until that time the fo'c'stle had been occupied at Pin Mill by old Alf whose portrait still hangs in the Butt & Oyster at Pin Mill. We spent the winter converting CENTAUR for the charter work having first renewed some inwale on the starboard side. The galley as originally put in included the mahogany worktop which extended right across the barge. There was a Barge Club connection as the worktop came out of the PRETORIA which John Prime and possibly others had bought when she was towed round to Maldon. The barge had been laying at Pimlico for years and the work top was part of the very plush fittings below. I believe the barge had been towed round by the guys who had rigged the CIV, mostly APCM lightermen, like Vic. Wadhams and the fittings were all dismantled by a largely Barge Club team which included Vic, Peter Love and myself.   Most of the T &G panelling and bunks in CENTAUR were installed by John Prime and myself whilst Tommy worked on the engine etc. At that time I had never had any experience of engines having always sailed with the Club since the Autumn of 1964 in the ASPHODEL with Vic Wadhams, Chris Dickenson and Harold House.

The engine certainly made a difference to our Maldon charter work and I believe that we were the only one to be fitted with at least a reliable engine at that time. KITTY was engineless as later on was DAWN although MARJORIE had a form of propulsion which was supposed to be jet propelled but was less than efficient. The engine did mean that we could get out of Maldon without the aid of a tug more or less regardless of wind and especially during the calmer summer weather we could undertake much longer weekend sails without using the engine but with the added confidence of knowing that we could still get our charterers back on Sunday.

CENTAUR was an extremely strongly built barge and I remember when we were working on the inwale together that Tommy said that her inwales were as large of those of REDOUBTABLE, a much bigger barge. Tommy had been skipper of the REDOUBTABLE for years as a motor barge and knew her strengths and weaknesses. Certainly in the time that I was in her as mate with Tommy until he retired we never seriously shipped the hand pumps. Any small amount of water which the barge made, and of course there was some always, was dealt with by the bilge pump connected to the engine as we motored in or out of Maldon where necessary. Even on the down channel work which we did later for the Bells advertising job we never had any cause to ship the hand pumps.

We worked the whole winter at fitting out CENTAUR. I was working in London and John Prime, of course, was also in full time employment elsewhere so Tommy would work during the day and then John and I would come down after tea around 7.30pm and work until about 10.00 p.m. every evening on the slow conversion into a charter barge. Then we would have a cup of tea in the cabin with Tommy and listen to some of his endless supply of yarns and anecdotes about the sailing days in barges and then we would go home and start again the next evening. Her gear was sensible in size and it made a very comfortable barge to sail and steer. As the Club have, no doubt, discovered, she never achieved any amazing racing form notwithstanding the best endeavours of Tommy Baker whose credentials in that respect were impeccable. You may wish to check Fred Cooper's *Racing Sailormen* to see his efforts in the 1930's with GENESTA which speak for themselves. Tommy liked to bend all the gear on when the mast and sprit were completely lowered. I never knew him to bend a topsail on after heaving up for example, unless, of course, the topsail had been sent down for repair. We never had any frills like an additional purchase to heave out the topsail sheet. The mate was expected to snatch the last bit of topsail sheet out whilst she winded. We always carried two 'handy-billy' tackles which were used frequently on race days to control the staysail sheets, We kept two chain 'snorters' in the shelf inside the cabin scuttle to attach them with. They also came in very handy down channel with only the two of us on board to heave in the mainsheet etc.

When we sailed the first year in the charter work we used the extra barrels on her big brail winch to the full, with wire foresail as well as topsail halyards. I had always been used to rope halyards on ASPHODEL and WESTMORELAND and I soon found that such a system has its disadvantages as well as advantages. With a single part wire foresail halyard it's harder to make the riding light fast to the wire. Also you had to be careful not to heave the topsail up too high, especially when setting it on the starboard tack. Sometimes you would discover that, just when you needed the sail down in a hurry, that you couldn't get the pawl. We evolved a system of sailing on the brake after slacking back a bit, and with the pawl down but not engaged. Tommy never shouted or even raised his voice but would just gesture with a finger when he wanted the topsail down, so you felt a bit stupid if you couldn't do what he wanted when he wanted it. Next year we went back to rope foresail halyards, the winch was too efficient and didn't give the charterers enough ropes to pull on. We always kept the wire topsail halyards though, the big winch also made it easier to get the topmast up, of course.

In spite of CENTAUR being such a slow-coach, races were always taken very seriously. We always took the propeller off the evening before. You had to be quick at sail trimming to satisfy Tommy and of course in light airs you had to be quick but *not jerk at it!* I remember one year, I think 1968, in the Blackwater we won our class (we would have been first overall but they shortened the course for the MAY when we were already below the new mark).

Capt. Tommy Baker, Sally and Eddie Smith

It was practically a calm all day, and of course it takes a lot of skill to sail a barge in conditions like that. Tommy was a non-smoker but it was comical to see him grab a fag out of someones packet and light up - so that he could blow smoke in the air to get a better idea of the direction of the tiny puffs of wind.

CENTAUR was a good sea boat too. I remember a hectic night when we scuttled between Cowes and Chichester with a heavy swell. We sailed stumpy rigged with the engine running and you had to be smart to get the weather leeboard up. Nevertheless we were dry enough. I watched an amusing conversation when we were alongside at Chichester. An old boy came down to visit on a bike, wearing old-time seamans rig of blue serge trousers, reefer jacket, peaked cap etc. He proudly announced that he was ninety something years old and had been in schooners for years. He and Tommy yarned for quite a while as the boozing went on below and Tommy. then aged 70, kept calling him 'Dad'!.

Tommy was very well respected by most other skippers. I overheard a fascinating discussion between him and Harold House once on the subject of mainsail headropes. The ASPHODEL had a new mainsail with one of those newfangled wire headropes. Harold had only ever had hemp as he'd been lightering for APCM since the '30's, so he came over when ASPHODEL was alongside at Maldon to ask Tom's advice about wire headropes and how best to stretch the new sail out. Tommy returned the compliment later, I was pestering him for information about how to use a setting boom in a narrow creek, after explaining he said "But the real expert is Harold House, you should ask him."

I sailed as mate with Tommy every weekend the first charter season and then the following year Peter Little 'joined the team'. Peter at that time was working for Architects Building Design Partnership as a model maker and you can see his skilled woodworking efforts in the large saloon table of the CENTAUR which he made from two smaller tables joined together. Next season Peter and I shared the job as mate with Tommy, taking more or less every another weekend in turn."

A brochure now in the archives of the Club shows that in 1967 she was sailing under the ownership of The Centaur Barge Chartering Co. Ltd. a company whose office was at 42 North Street, Maldon. They chartered CENTAUR to holidaymakers at weekend rate of £54 for April and October, £60 for May June July and September weekends with weekly rates ranging from £82 for 8 up to £110 for parties of 12. Charter rates did not include food which was supplied and prepared by the parties themselves. Blankets pillows and mattresses were provided but charterers were expected to provide their own sheets.

She entered the Whitsun barge matches at Southend and Medway in 1965, only a few months after she had been bought out of trade in her unrigged condition so it was hardly surprisingly her race record was not immediately outstanding. CENTAUR was one of the 11 entries for the Southend Match on 5th June 1965 and the Medway Match on 7th June She also sailed in the Orwell where she followed home EDITH MAY, MAY, WESTMORELAND, KITTY and VENTURE.

In the Second Medway Match on 28th May 1966 13 barges were entered and whether the crews were superstitious or not the event was certainly not without incident. The weather was dull and cloudy with a north east force 4 and the BBC Television cameraman in a 36 ft open boat looked as if he had been dropped in a bucket of water. First across the line after the gun was CENTAUR so winning the Challenge Cup presented by J.P.Knight Ltd. While the MAY swept through the field

CENTAUR was in collision with C.I.V. resulting in a smashed boat in her davits. The Club had entered WESTMORELAND but her crosstrees bent letting her backstay slacken causing the topmast to snap so putting her out of the match. CENTAUR came in 8th out of the ten craft racing that time but Mick Lungley was first over the starting line again in the next match, at Southend on Whitmonday 30th May when 9 barges started. This time he was presented with a silver tankard plus a half bottle of scotch from the famous company Bells who also presented similar prizes to other winners but finished 7th.

She was again 7th out of 9 in the Blackwater on 9th July. Again thirteen barges started in the Pin Mill match on 13th August when CENTAUR was beaten to the line by Spinaway C. This was the first match to be held after the Darwin disaster and the Board of Trade were threatening to introduce a 12 passenger limit. Several barges decided to withdraw from the race but the organisers got the agreement of the Board of Trade that the match could be held within "Smooth Water Limits" with the outer mark as the Cliff Foot Buoy, off Landguard Point and the whole course in the Orwell and Mistley Rivers. This course which was used for several years and only in recent years have the fleet sailed out to the Cork. Despite all the difficulties it was a most successful race with CENTAUR finishing in the middle of the field.

In the following year, 1967, the races were fairly uneventful for CENTAUR save that ARDEER beat her by a tenth of a second. She retired in the Southend match on 27th May, was 6th out of 11 in the Medway and 9th out of 12 in the Blackwater on 31st August. The next year, 1968 she was 8th out of 10 in the Southend match on 1st June, and 5th out of 9 in the Medway on 3rd June. Richard Duke was again dealing in barges; he sold MILLIE and bought CONVOY which had been squashed in Tilbury Docks. He was right in thinking that she would return to shape when a bent steel beam was taken out and he had her quickly converted back to sail at Pin Mill with sailing gear from LESLIE WEST and a mulie mizzen made from the topmast of LORD CHURCHILL. CENTAUR gained her first first place in the Blackwater match. being the first barge home in Group One in the drifting Match held on 13th July and took the championship flag, cups for the first over the line and the B.S.B.M. cup for 1st at the outer mark. Sailed in light airs and skippered by Tommy Baker she beat WESTMORELAND and CONVOY.

For many who sailed on charter parties at this time it was a one off holiday but for some, like Lindsey Simon, (then Lindsey Day) it became a way of life. She has written from her home near Saxmundham that "It was a fine summers day when I saw my first sailing barge on the Blackwater and I resolved to sail on one somehow. The opportunity arose in 1968 when I heard about a charter being organised by Valerie Smith on board CENTAUR for a week, the week that included the Southend and the Medway Matches.

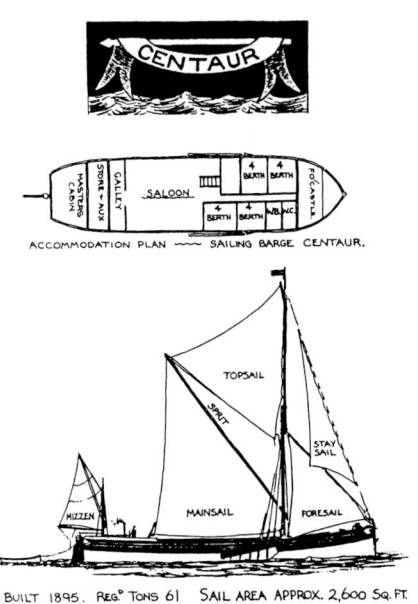

ACCOMMODATION PLAN ~~~ SAILING BARGE CENTAUR.

BUILT 1895. REG.º TONS 61	SAIL AREA APPROX. 2,600 SQ. FT.
LENGTH 82 FT.	DECK TO TOPMST TRUCK 75 FT.
BEAM 21 FT.	SPRIT 60 FT.
DRAFT 4 FT.	FULL STANDING HEADROOM.

CENTAUR
BARGE CHARTERING
CO. LTD.

Centaur Barge Chartering Co.Ltd.Brochure

The charterers were to be met by the mate, Peter Little at the end of Southend Pier and ferried to the barge lying at anchor. We rolled badly all night, the one and only time I have ever felt seasick on board. However, in the bright light of the early morning the sight of all the barges getting under way soon eradicated thoughts of lack of sealegs. It was a beautiful day with a good breeze. Although, as I later realised, CENTAUR was a slow old box, Tommy Baker the skipper took the race very seriously, particularly the start. "First over the line" was the only prize we had a chance of winning! Tom's toothless gums were working away nineteen to the dozen while he concentrated on getting a good start. Tom's ill-fitting set of false teeth lived in a cupboard in the Master's cabin, he never wore them. In the years that followed I became a dab hand at cooking his fried eggs so that they weren't in any way chewy. The KITTY I remember had a minor collision with the MAY while approaching the start line which added to the general excitment.

The Medway Match was held on the Monday and the start of this race was even more exciting, it being so congested on the river. KITTY kept well out of the way this time. Though we started well, we couldn't keep up with the fast barges. But we enjoyed the sail all the same. Though Tom was a mild natured man, he could be irritated by bad manners and sailing incompetence. While still racing and turning all the way back up the Medway we were overhauled by the sail training yacht DUET which was motor sailing. Every time we came about she was in our way. Tom strolled over to the rails aft and spoke in a loud clipped voice...."You are still the overtaking vessel....so keep out of my bloody way."

This first holiday aboard a barge was memorable for the good company and variety of places visited. I was totally hooked. It was the first of many super sails I had with Tom, of whom I grew very fond. He had led an extraordinary life at sea, in particular during the war when he joined the Royal Engineers Inland Water Transport. As an experienced Sailorman he quickly proved his worth and in due course was in command of a tug in the Adriatic. He was a brave man and was awarded the MBE, but very modest about his experiences. He was proud of one thing though: At the end of the war he was challenged by an Italian Colonel to have a go on a gondola. He watched these being propelled and worked out how to do it, and without too much trouble he (in his own words) "made a good job of it."

Later in 1968 CENTAUR won her class in the Blackwater Sailing Barge Match. Rumour had it that this was because her bottom was rough and she managed to hang on to the water and not drift astern like all the others. The crew were Tom Baker, Eddie Smith, Peter Little and John Prime. Having been bitten with the sailing bug, I volunteered to help with tarring round, painting etc. and in due course sailed with Tom and Peter as third hand when the barge was chartered.

As any crew will tell you, the first night of a charter party is a noisy, tiresome affair: These 'adults' are full of excitement, to say nothing of beer. We would sail from Maldon Quay on the Friday evening, endeavouring to keep all drunken charterers from falling overboard. They very helpfully would bring their torches up on deck, which only served to blind those of us with 'night eyes'. Tom was not very tall, and they would gather around him at the wheel. In desperation he was often heard to ask "Is your father a glazier?", to which the puzzled charterer replied "No". Tom would rejoin, "Then that's why I can't see through you!". After Saturday morning hang overs, most charterers settled down and usually we were all firm friends.

Tom's wife became ill later that summer and he left CENTAUR temporarily. Theo Horlock (Chubb's cousin) took over for a while. Theo was quite a different kettle of fish. He wore a long raincoat and plimsolls and was content to 'twiddle' the wheel and yarn and let you get on with everything else - "Yew do jus wha yew warnt...." was his favourite expression. Sadly, Tom's wife died and in due course he returned to Maldon to take over from Theo."

Eddie Smith recalls that "Tommy had done the Bells advertising job with Barry Pearce as mate in LRTC's SIRDAR and he maintained his connection with the PR firm which arranged everything for Bells when he moved into CENTAUR. When Tommy secured the Bells Advertising job for CENTAUR, both Peter Little and I spent all our summer holidays on the Bells job down channel. I seem to remember in the summer of 1969 Peter was mate from London to Torquay, I did the leg from Torquay to the Isle of Wight and Barry Jenkinson did the final leg with Tom from the

Isle of Wight back to London. The only additional equipment which we needed for the trip down channel for Bells was additional heavier synthetic fibre mooring ropes which were something of an innovation at the time, at least for sailing barges. The schedule of receptions, booze-ups etc. organised by Bells at time was not quite so hectic as it has now become so that probably 75% of our trip was done under sail.

The Bells Scotch Whisky Charter 1969   *Fotoflite, New Romney*

The consequences of one of those Bells trips were still visible when the Club took over many years later in the form of a longer topsail headstick than she originally had. We had arranged to pick up the public relations lady for Bells from Ryde Pier during one of these trips but the weather had turned extremely blowy with torrential rain and thunder. During one heavy squall and accompanied by a loud clap of thunder the topmast was struck by lightning. By this time needless to say we had decided that nobody would be interested in a days sailing from Ryde Pier so we anchored to try and ascertain the damage. After studying the topmast pole through binoculars Tommy came to the conclusion that only the actual pole itself, above the eyes of the rigging, had been shattered and that it was still safe to set the topsail. We were by this time on our way back to London and after our return Tommy resolved the problem of the shattered topmast pole by simply reducing the length of the top mast by a few feet to accommodate a new pole from the undamaged wood, shifted the rigging down and then that winter we brought the topsail down into the saloon and Tommy recut the head of the sail to accommodate a longer headstick and thus compensate for the shorter luff length of the reduced top mast "

Lindsey Simon takes up the story again in the following year 1970, writing that CENTAUR "was chartered again and this time I accompanied Tom and Peter from Maldon to Falmouth. We arrived at Cowes, Isle of Wight at the beginning of Cowes Week and Tom knew exactly where to anchor to get a good view of the Royal Yacht BRITANNIA and the Trinity House Vessel PATRICIA but be out of the way of the noisy hovercraft ferries. Both these vessels were dressed overall during the day and lit up at night and it was quite a sight. We dutifully sailed the barge about so all and sundry could see our sails, then continued our voyage.

At Weymouth we entered the harbour under full sail, rounding round, brailing up and coming alongside as neat as a pin. I remember Tom being rather cross with me for grabbing a fender in case we should need it. Of course, we didn't. It was necessary to motor sail some of the way in order to try to get to the destination according to Bells' timetable. They rather expected us to arrive like a train, irrespective of tides and weather. We did meet some rough weather crossing Lyme Bay and despite being lashed, the anchor clonked about all night, and the sound of the bows crashing into the seas made the foc'sle an uncomfortable place to be. We were very well aware that the old barge should be nursed along to some degree and, quite rightly, Tom tried not to push her too hard. I can't quite remember if we missed any receptions because of the weather, fortunately we were out at sea and there were no mobile phones in those days. We arrived safely in Plymouth and tied up at the Pilgrim Steps where we were welcomed by the "Pilgrim Fathers' all dressed up, it being 350 years since the MAYFLOWER left Plymouth from that very quayside. Regimental Sgt. Major Britton, acting Town Crier in his red outfit came on board for the reception for local publicans and dignitaries. Our cargo of whisky was consumed enthusiastically in the hold which was lined out with tartan cloth and our task was to ensure all visitors were returned to land safely.

Centaur in Plymouth, 1970

We went to the precaution of erecting stanchions and ropes all round the rails to be sure. While at Plymouth we had a visit from a retired doctor who brought with him a model of a barge he had made. Amazingly it was a model of the CENTAUR. He had had a trip or two on her when she was in trade and he was a medical student in London,. He was as amazed to see this very barge again as we were to see his lovely little model. At last we got underway to escape the crowds and sailed up and down Plymouth Sound in full view of the spectators on the Hoe. Our furthest port of call was Falmouth and it was there that I left to return home via British Rail.

Eddie Smith joined CENTAUR for the return voyage taking over from Peter again. On her return to the Thames, Tommy brought CENTAUR up to Tower Pier where there was yet another reception and Sgt.Major Britton put in another appearance. The charter was completed and Peter and I rejoined the barge and together with Eddie Smith, Rick and Fay Hogben and Maggie Nobbs (a regular mate on KITTY), we got under way early in the morning. Tower Bridge lifted for us and we set off down the London river in the mist before dawn broke. We passed the CAMBRIA moored below Greenwich and Bob Roberts gave us a wave, and as the sun broke through the mist, Tom lowered the gear and we all took much delight in cutting off the "Bells Scotch Whisky" letters. Peter dressed up in tartan to help celebrate which reduced Tom to tears of laughter, which was my favourite memory of all."

Tom Baker, with Peter Little in "kilt" & bottle   1970
Cutting the letters off the Mainsail at the end of the Charter

56

CENTAUR sailed back to Pin Mill hard and Rick recalls that it poured with rain with not much wind, but Richard Duke gave them a splendid dinner on LESLIE WEST. The following year Bells acquired the HYDROGEN for their promotions and took her further afield. For many years a photograph of CENTAUR under Tower Bridge taken on the charter hung in her saloon.

When Gordon Wright took over DAWN as skipper in 1970 or thereabouts, Peter Little went mate with him and Lindsey has added "Peter and I married in December 1970 and Eddie was Best Man. Tom came too, having first 'reccied' the venue the week before so as to be sure of finding the church. For the first five months of our marriage we lived on the SALTCOTE BELLE. Peter was a shipwright and I worked at Dixon Kerly's boatyard and then joined Barry Pearce at Walter Cooks where I stayed until the bargeyard was sold to Roger Beckett in 1981. Peter was keen on steam engines and when Barry bought his steam tug CHRIANIE, it rather took over our lives."

CENTAUR was again the first barge home in Class B in the Blackwater Match on 11th July 1970 followed by VICUNIA and SALTCOTE BELLE in a moderate westerly breeze. The Southend match, held on 29th May, 1971 was another notable triumph for Tommy Baker sailing under Richard Duke's house flag when he was first over the line at 09.00 and first to finish at 12.12 closely followed by KITTY. The match was sailed in ideal conditions with a force 4 south west breeze.

After leaving CENTAUR, Eddie was mate with Bob Wells in MAY and in 1972 he went back, once again, to office pen-pushing spending only a few more weekends on CENTAUR with Tommy Baker. In October 1972 Tommy Baker decided to retire and CENTAUR was laid up at Pin Mill while Eddie Smith went off to spend the couple of years in the Far East. When he returned she was looking somewhat sorry for herself, her stem, frames and planking showing signs of incipient rot which would soon need treatment if this was not to be her final resting place as it was for the many of the houseboats and hulks which were drawn up on the shingle beach under the trees on the shoreline . Tommy Baker had bought a little 'Dauntless' the "DONNA" which was a regular visitor to Maldon until his death in 1985.

Lindsey and Peter Little eventually parted company; Peter became skipper of the MV SAND KITE living in Oswestry while Lindsey moved with her new husband to farm in Suffolk. CENTAUR's owner, Richard Duke finally sold CONVOY in 1978 after ten years of sailing her. He continued to follow the matches in his Scottish fishing boat ADAX and finally in the Ipswich fishing boat ANNIE D but had many other interests including windmill restoration having in1985 bought the derelict Stanton Post Mill which he had turning again just four years later.

Capt. Duke and his wife Annie remained 'at home' on LESLIE WEST to his many sailing friends including a rugby playing representative of Duckhams Oils of the name of Barrie Lawrence from Grays, Essex, who had become a leading member of the Thames Barge Sailing Club. They met up again on 6th March, 1993 at the "Bargemans Ball' organised jointly by the Sailing Barge Association and the Association of Bargemen on H.M.S. BELFAST at her permanent berth in the Upper Pool along with many of old friends including members of the Pin Mill Barge Match. Only a few days later on 20th April, 1993, Richard Denis Duke died at his home in Lower Holbrook after a short illness aged 79 years. His friends came together in respect for a Memorial Service at St.Mary's, Harkstead on Saturday 8th May, 1993. So ended another another link in the chain of the life of CENTAUR.

Barrie Lawrence, President Thames Barge Sailing Club

5. The Club buys a barge.
1974 - 1977

"Club formed to save the Thames Barge" ran the headline in the Observer for 18th April, 1948. The meeting of a few enthusiasts at the National Maritime Museum on 12th November,1947 had led to the adoption on 6th March, 1948 of a formal constitution to preserve and maintain in commission one or more Thames spritsail barges. Hugh Vaudrey had been the inspiration behind the idea, burst into Frank Carr's office at the Museum to announce his scheme and added, "and _You_ are to be the first Commodore". By the end of the year the membership stood at 145, many having connections with the Greenwich/ Blackheath area.

The Club chartered SPURGEON for four months and ran a weekend shuttle between Greenwich and Whitewall Creek, opposite Chatham at a cost of 12s 6d for Maritime Members and 17s 6d for Associates or Guests. In those days members were sent a telegram offering a berth for a weekend and asking them to report at where ever the Barge might be and to "signal acceptance to Hugh Vaudrey". From the outset there was a concern to collect information relating to the history of the craft and a card index and photograph albums were compiled and these remain in the Club archive. By 1952 a monthly Bulletin, was being sent to members to keep them in touch with changes in the barge fleet and club activities including meetings or 'musters'. In the January Bulletin there was a note that Stan Yeates had come to a muster direct from the Surrey Docks where CENTAUR was loading wheat for Felixstowe on the passage he described in Chapter 3.

Eastwoods Ltd. gave WESTMORELAND to the Club in 1963 in the hope that she could keep sailing for just another couple of years. Vic Wadhams, then working on A.P.C.M. tugs became an occasional skipper while the regular skippers were Bill Wickenden, Bill Hughes and until the end of 1969, Harold Farrington-House. There were still barges at work in their traditional trades but they were steadily being superceded and in 1968 London & Rochester Trading Co. Ltd. decided to dispose of PUDGE. The Club made a rapid decision to buy her and that winter members worked to convert and re-rig her: the Club had now really started on the task of barge preservation, as well as just barge sailing. Late in 1970, through the good offices of the then Commodore Geoff. Collard of Silvertown Services we acquired one of their lighters, now named SAILORMAN for use as a floating store/ workshop and she remains an invaluable member of the fleet.

In the Club Bulletin issue 241 the Commodore Harold Eagles had the unpleasant duty of informing the members that WESTMORELAND had been lost. On Saturday 18th August 1973 she became stranded at Hoo with her forefoot on the concrete mooring lighter. Just after midnight her bows broke open, the stem pulled out and the foredeck lifted. She flooded on the next tide and sank. It was beyond the resources of the Club to salvage her and the extensive damage probably put her beyond economic repair. All the gear which could be saved was taken to Faversham for storage and the Club claimed against the Underwriters. A sub committee was set up under the chairmanship of Walter Swanson, a past Commodore to report and advise on the feasibility of repair to or the acquisition of another barge hull. The club asked for estimates from yards at Hoo and Maldon but these were between £10-£15,000 while London & Rochester's Acorn Yard considered she was too badly damaged for a repair to be considered.

Barrie Lawrence succeeded Harold Eagles as Commodore at the Annual General Meeting on 30th March 1974 and at the meeting of the committee on 7th May the new Commodore reported that Richard Duke had offered CENTAUR to the Club. Captain Fred Cooper had inspected the barge and reported favourably on her condition and a survey by John Perryman showed that she was still seaworthy and all her gear was intact although needing some attention. She still had a full suit of sails that with a few patches would suffice for a season or two, what was more, she was completely converted to accommodate fourteen passengers. Roger Martyn, a solicitor member who was one of the Trustees of the Club offered to conduct negotiations with Richard Duke's solicitors who were offered £3,000 down and the balance to be paid within a period to be agreed. They requested £6,000 down which was agreed and the rest on a three year interest free loan. All we had to do was raise the £6,000, give her a good overhaul and put her to work to earn the rest of the sale price.

Gwen Lawrence came up with a scheme for selling "CENTAUR Bonds" to members and Roger Martyn drafted the form which Peter Love arranged to have printed without delay. The idea was that members would buy £5 Bonds and that they would have a chance in a monthly draw when the lucky member would get back £6. So, the Committee made the decision to buy CENTAUR. Enough Bonds were sold, helped out by a few donations and a loan from Barrie to pay the full £6,000 and he completed the purchase with Richard Duke. So the Club took over CENTAUR in July, 1974.

CENTAUR had lain under the trees at Pin Mill for just over a year without sailing before we bought her. Terry O'Sullivan has written " My first visit aboard her was with a small working party in August 1974 to check over the gear and tidy up prior to her first sail. John Dearden was Rear Commodore and arranged for her to be moved from under the trees to the bottom of the hard where we could clean her up. On this occasion there was just a few of us, John, Mike Taunton, Fred Cooper, Barrie Lawrence and myself.

We struck the topmast and John climbed on to the hounds and fitted the Club bob for the first time. The top mast was hove up again and we proceeded as best we could to check over the gear. My input here was, I must say, minimal although enthusiastic as I was a newish member and still green. My impression of CENTAUR was that she was less boxy than PUDGE but of course much heavier than WESTMORELAND and I doubted she could ever give such a sail as my last one in her in a gale. Down below all the panelling was impressive and the skippers cabin was a delight to behold. A week or so later I was fortunate to be on the first sail with the Club."

After a few weekends spent at Pin Mill cleaning her up and rigging out, a crew was assembled and she was sailed to Faversham by Mick Lungley who wrote "Enjoying some shore leave trying to forget about the sea and water, replacing a pane of glass in a bedroom window, a car ground to a halt in the road. "Hello Mate" the driver calls out in a Grays cockney accent. It was none other than, our Commodore, Barrie Lawrence. "thought I would give you a look; been down to Pin Mill, heard you were on leave, wish I'd known earlier, wanted a skipper to take CENTAUR over to Faversham" He said all that in one breath, then, over a cup of tea, told me that Vic Wadhams would be coming over on Monday to take her there but Vic was very busy with a weekend on the PUDGE. I said if the Club were pushed Barrie should give me a ring.

On Saturday morning I got the telephone call, Vic would be pleased if I could help him out. The crew would be on board on Monday evening. In the afternoon I went aboard CENTAUR to see where the mooring ropes were leading to (it would be dark when we cast off the next morning.) In the evening I went into the "Butt and Oyster" to meet the crew of John Dearden, Colin Frake, Brian Weaver, Peter Maxey, Jim Gregory, Terry O'Sullivan, Cliff Manning and Colin Fox (Barrie Lawrence had to miss the trip due to business commitments). A likely place for everyone to meet I thought. After three or four pints of Dr.Watts early morning rising bitter I told them I would be aboard at 04.30 hours on Tuesday 8th October, 1974.

At 04.45 Colin Fox and myself mustered aboard Centaur; not a sign of movement from the crew. We had to wake them up because they had no alarm clock but it was not long before everyone was about. It was a calm, drizzling morning, wind about S.E.. 0-1, the engine was started and the ropes cast off and we were under way at 05.30 hours. Motoring down through Butterman's Bay we had a cup of tea and then hove the boat up in the davits.

As it was coming daylight we set the topsail, mainsail and foresail and shipped the rolling vangs while we were inside Harwich Harbour. At 06.25 hours we passed the Beachend Buoy, wind easterly force 2, sails just filling but we still had the iron topsail. Twenty five minutes later the Stone Banks Buoy abeam there was much activity on deck as the crew tried to get Pin Mill mud washed off the decks. By 0800 hours the wind had moved into the north about force 3. We gybed and set the staysail with the smell of bacon frying coming from the galley and it was soon followed by the call of "Breakfast"

At 09.15 the Wallet Spitway buoy was sighted in a heavy shower of rain and half an hour later the engine was stopped as we passed it. Wind was north force 3 and we dropped the staysail, picked up the skirt of the mainsail and gybed. Then we dropped the mainsail out again and held this gybe for about fifteen minutes to clear the Whittaker Spit, when we picked up the skirt of the mainsail and gybed her back on the starboard tack. The wind had increased to Force 4 North and the old girl was coming to life again, she had not been under way for about two years, and we roared past the South Whittaker at 10.25 hours.

We were up to the south west Swin by 11.39 and it was not low water but rather than wait for the tide to make, so we could sail across the Red Sands, I suggested making a sail of it and go down to the Girdler Buoy and then past the Whitstable Street. The crew were all for more sailing. We soon brailed the mainsail, dropped the head of the topsail and gybed her on to the port tack. Topsail was reset again, mainsail sheeted in, the crew all working well and saying what a good day for a sail. The wind was abeam now and a bit of a swell was lifting her chines out of the water and 12.25 we passed the Shivering Sands Towers and in ten minutes were abeam of the Girdler Buoy and it was time for another gybe. By this time the crew had gybing off to a fine art. It was down topsail, up mainsail, catch a turn round the toe with the mainsheet, slack the backstay, stand by the lee vang ready to take it in and she gybed over lightly onto the starboard tack. Sails were set again and we passed the Whitstable Street Buoy and as we came into the East Swale past the Ness Houses the wind was freshening to Force 5. We stowed the mainsail at 14.20, dropped the foresail and five minutes later dropped anchor and as she swung on the flood the topsail was dropped. and clewed in.

We had to wait for the tide until 15.20 hours we started the iron horse, hove up and proceeded up Faversham Creek. At 16.15 CENTAUR was moored alongside Gillets Wharf at Faversham. Everyone was pleased with the days sail. I know I was, it must have been one of the best passages made this season. Thanks to Colin Frake and British Rail I was back in the Butt and Oyster before closing time."

The first mention of CENTAUR in the Bulletin is in issue No. 251 where there is an application for the Bonds, the first of which was issued to one J.P.Boss, (of whom more will be heard later). CENTAUR was brought to Greenwich to be shown off to members and public alike at an Open Day exhibition. This was held with the object of making money for the CENTAUR fund and so club members had to pay an entrance fee the same way as the general public The venue was at Greenwich pier, the date 26th/27th October 1974 and the admission charge adults 10p, children 5p. As well as PUDGE we had MAY belonging to Silvertown Services Ltd. and the World Trade Centre's DANNEBROG alongside the pier with an exhibition of models, paintings and photographs on PUDGE and MAY with DANNEBROG used for film shows.

With Pudge and May on Greenwich Pier for the Open Days, 1974

The star of the show herself left Faversham just before nine on the morning of Friday 25th October 1974 under Capt. Harry Parter with Peter Maxey mate and a local Kentish crew. They motored up through the Swale, past Garrison and through the Swatchway using over 25 gallons of fuel to arrive at the Pier at nine thirty that evening. The crew returned home late that evening and their successors who were to man the barge for the Open Day came aboard. That night a wonderous pine fragrance infused the cabins: the stove in the saloon had overheated and started to burn through the ceiling! The event was nevertheless counted a success with about £350 made towards which by the following month included £1,435 raised through the Bonds, gifts £746 and £1,310 in loans.

In the evening the club held a supper onboard the Cutty Sark which Rolf Harris the showbis personality attended. Jack Jeffs donated 12 superb 15"x12" photos to be sold in aid of funds and Peter Ferguson produced greeting cards for Christmas or birthdays. CENTAUR had no wash basins in the cabins and these were considered essential along with galley utensils not to mention paint, rope wood etc. for use during the winter. That winter while PUDGE was due to go into Junction Basin in the West India Dock, CENTAUR returned to Faversham, sailing away from the Open Days at eleven on the Sunday morning in a good 5 - 6 North Westerly.

 She went a short way upstream, turned, and still with a hint of bunting on her topmast stay set off downstream under full plain sail, a magnificent sight for those on the Pier. It may not have been such a good journey back as the log ends at three in the afternoon with her in Sea Reach facing a force 7/8, but she duly reached Faversham where Colin Frake took charge of the winter working parties.

After her first official appearance for the Club CENTAUR spent the winter at Faversham where there was a good experienced crew of members and that is where the first part of her restoration began. They say that to do a job properly you have to start at one end and keep going until you get to the other. What better then that start with the rudder, which was in in a pretty bad condition. The Faversham lads including Colin Frake, Brian Weaver, Peter Maxey and Geoff. Harris took the rudder from the wrecked WESTMORELAND, extended the length of the blade and hung it on CENTAUR. They also repaired some areas of the rail. The gear was overhauled, sails patched, the engine serviced and she was put back into commission in 1975 as the Clubs' charter barge.

The charter fee was £200 per week, skipper mate and the use of all equipment inclusive and Club members were offered a 10% reduction but the charterers had to supply enough food for themselves, the skipper and mate. The Charters were organised by our Sailing Secretary, Colin Welsby and the first charter was by the Sou-West Sailing Club from 17th to 23rd May 1975. Vic Wadhams with David Ward as mate and Alan Phillips and friends from the West London sailing club took her to Maldon, up the Thames to Gravesend, then up the Medway to Rochester Bridge before leaving her alongside the lighters at Hoo. Vic stayed onboard to take her away again for another weeks charter which included 2 days windbound at

Hoo and on 31st May 1975, the Medway Match.   She finished 9th in her class
behind PUDGE but ahead of REMERCIE and EDITH MAY.   The Faversham lads
had their reward with the chance of sailing her over on a NALGO charter to the
Blackwater for the match on the weekend 14th June 1975 when under Vic
Wadhams she finished in ten and a half hours ahead of ARDWINA, PUDGE,
KIMBERLEY, SEAGULL II and REDOUBTABLE.   The log was left with a note for
Gordon Chell, the next duty mate, that a small repair was required to her foresail
where a tear had been caused by a shackle on the staysail sheet, also he should
seize the peak brail blocks on head rope.   The weekend duty mate hoped Gordon
and his crew had a good weekend and the following charter under Alf. Larraman
included the Pin Mill Match sailed on 28th June.   CENTAUR was first over the line
in class B at 08.30 finishing at 14.15.

Terry Everett brought her back to Hoo for the Norfolk Wherry Trust charter on
25/27th July, and she remained in Kent until the Third Swale match which was
sailed on 16th August, 1975, the weekend before the Southend Match.   Terry
O'Sullivan had put his name down for crew and looked forward to taking part in her
first appearance in the match. He reported, "We joined CENTAUR at Faversham,
the skipper was Harold House and the mate Fred Woodward and motored down
the creek to anchor prior to the start.   The early Swale matches were anchor starts
and could prove to be lively. On this day we got away safely and sailed outward to
the Spaniard.   As we approached the Colombine Buoy we were slowly being
caught by CIV and Harold took us as close to the buoy as possible to prevent the
CIV overtaking us.   Her skipper shouted over "let us through mate" but Harold
replied "We're racing don't you know".   CIV's skipper tried his luck but when it was
evident that she couldn't get past us completely before the buoy, he had to bear
away.   As he did the mizzen sheet flapped and as luck would have it caught around
our port stern snatch.   As the two barges parted the sheet became bar tight until the
mizzen sprit collapsed and the sheet broke, but not before bending the after davit
and sinking CIV's barge boat which then bent the forward davit.   The language
used at us by the female members of the CIV's crew surprised us. We were sorry to
see the damage done but felt the skipper shouldn't have tried to overtake us on that
side, but that is all part of racing.

One of the barges to slowly overtake us was the ETHEL MAUD, on what proved to
be her last Match outing. Harold had us all in stitches with his constant dry humour,
with the quips coming so fast. They say trouble comes in threes and this day
proved it so. On the return trip we were cross tacking with the smack ROSE AND
ADA in the narrow channel off Shell Ness. Quick reactions were required here
because as one of us went about the other had to follow quickly. Also, because of
the shallows our leeboard had to be raised quickly. Unfortunately, on one tack the
leeboard was let down too far and bottomed which slowed our coming about. The

next thing we knew the bowsprit of the ROSE and ADA punched a hole through our starboard rail just in front of the main horse. The bob stay acted like a saw cutting into the rail and as the vessels drifted apart one of the smacks jib halyards became bar tight, fearing a repetition of the problem with CIV's mizzen I quickly cut the halyard and was sworn at by the smacks crew for my troubles!

Our third and final catastrophe that day was after passing over the finishing line. Harold sailed us between two barges but underestimated our leeway and we cracked into the anchored ETHEL MAUD.   Her crew seeing the danger tried to lower her boat but only succeeded in dropping the stern completely . The boat instantly became a rather large but fragile fender which became matchwood as we started alongside. As we were still sailing the two barges cross trees locked and bent and those on the ETHEL MAUD tore our mainsail.   Not the way we had wanted to finish the match. Knowing the barge was needed for a charter the following week we decided to go back up the creek and lower down on the top of the tide. Despite this unfortunate accident,   Harold's spirit was never shaken and he sailed CENTAUR for us again without incident and his wit kept the crew always smiling."   Fred Woodward, the mate put a note in the log of a special vote of thanks to the crew, and to Harry Parter and Jim Gregory who came aboard to help to drop the gear and assist with the repairs on Sunday.

Tate and Lyle were going to Holland with MAY for 6 weeks in the summer to take part in a Dutch Barge Rally and they considered chartering CENTAUR to go with them but in the end it was decided that PUDGE would take her place.   PUDGE was a newer and stronger barge, and one of the weaknesses of CENTAUR, (which was already earmarked for the first available funds), was the condition of her stem and apron.   So our barges changed roles.   PUDGE became our charter barge and CENTAUR became our Club barge and was therefore able to take part in the Port of London Regatta Barge Match on Wednesday 27th August. 1975.

This was just one element in the programme of events organised by the Port Authority between 23rd and 30th August, 1975 supporting the 1975 Clipper Race from the Thames to Sydney which culminated in a procession of craft down river to Southend where the race would be started on Sunday at 09.30 by Edward Heath on a line between No.1 Sea Reach and the Medway Buoys.   At least four square riggers were expected to be in St. Katharines Yacht Haven with some 70 sail training craft in the river.   Geof. Collard was on the Steering Committee under the Chairmanship of Capt. Duggie Dear, the P.L.A. Harbourmaster.   Gravesend Council agreed to donate a trophy and pennant and some 24 barges took part in the match which was sponsored by the Financial Times. The barge match was officially organised by the Club with a start off a line from the Collier Signal Station Tilbury to the Customs House Pier at Gravesend.

CENTAUR sailed round from Faversham taking in the Southend Match on 23rd August and in the Spring 1980 Journal Martin Phillips writes "I was fortunate to be one of the crew on CENTAUR for a nine day club charter taking in both matches. It was the first time I had been to Faversham or sailed on CENTAUR for that matter. After meeting the others in a nearby pub we walked down the Iron Wharf tripping over the mast and spars lying on the wharf waiting to be fitted on the brigantine EYE OF THE WIND then fitting out at Faversham for a round the world voyage. Our skipper was Charles Frake, a well known Kentish barge man and among the crew were his son Colin and the 'Faversham Lads' who had been closely involved in rebuilding the club's previous barge WESTMORELAND.

We motored out of Faversham Creek bound for Southend. We spent the night motoring around the island heading back into the Thames estuary. We had a passenger on board who was joining the crew of ORINOCO a large barge which had been newly rerigged by Laurie Tester and which looked very distinctive with her green painted hull and white sails. It was touch and go if we were going to get to the start. Our passenger was dropped off, the sails hurredly set and our engine turned off. The gun fired and we made a splendid run over the line. Unfortunately we were the only barge that did go over because what we had heard was the five minute gun!. We had to beat back to the line and by the time we started most of our

class were well away. We ran out to sea and during the course of the next hour or two were were overtaken by the fast barges in brilliant sunshine. There was a thrilling finish to the match off the end of Southend Pier with MIROSA just scraping home ahead of ORINOCO. The match supper that night was a memorable occasion held at the 'Kursall' Ballroom. Next day we put into action a master plan to improve CENTAUR's image. On the after -noon tide we dried out off Leigh on Sea and all hands set to, scrubbing and scraping the sides of the hull. The following day the whole barge was repainted, sides blacked, the under water section painted red and the covering boards white and the name boards cut in in yellow and blue, even the transom

was repainted. By the time we floated off on Monday evening the tatty old CENTAUR was transformed. On Tuesday we sailed up to Gravesend. The decks were still being painted which posed some unusual problems in handling the gear. That evening we anchored above KITTY and GLADYS near the start line off the Ship & Lobster and went ashore. My memory may be faulty but I do not recall experiencing any difficulties going ashore. Times have changed because in recent years it had proved almost impossible to land there. I do remember that at some stage in the evening the two Colins, (Welsby and West) decided to lead the rest of us on a long trek in search of a pub serving better beer.

Next morning saw the start of the race to Cherry Garden Pier with a fleet of twenty five barges, the biggest fleet for many years. Certain things stick in my mind. The crew of LORD ROBERTS seemed to be permanently pumping. There was a full crew on board MAY who probably filmed our setting boom falling over the side just after the start line. ORINOCO made a start which somehow managed to put two reaches between her and the rest of us. There was also an element of doubt on board CENTAUR as to where the start line actually was. The departure of ORINOCO on a race of her own left us well us well out in front of the rest of the fleet and we were reunited with our setting boom which had been picked up by a passing boatman. Slowly we were overhauled by the faster barges which gave a great opportunity once again for photography. LADY DAPHNE looked superb, setting a huge red balloon stays'l.

Further up the river life becaume quite hectic with the hazard presented by various moored lighters. Use of engines was permitted to get out of danger but we were rather put out when JOCK motored past us in the course of our trying to avoid some lighters. This was the subject of some fighting talk on board when we finished third in our class instead of the second we deserved, so we wanted to protest. Unfortunately the lack of pen and paper on board meant that we failed to comply with the rules which required a written protest to be delivered to the Committee within a set time after the finish and nothing more was heard of our protest. The finish was itself spectacular. There were some massive square riggers moored in the river below Tower Bridge and fifteen barges arriving just below the bridge on a flood tide made great fun for all.

Eventually we all ended up moored on Butlers Wharf which was then a brewery, not the 'yuppie' development of flats or offices that it is now. The prize giving onboard VENTURE was interrupted by the owners' announcement that he was concerned that the dozens of people standing on the top of his main hatch were about to descent into the hold. CENTAUR finished in third place behind ORINOCO and JOCK with ETHEL in fourth place in Class C. Colin noted in the log that it was a very good race with very good crew who worked well together.

That night everyone scrambled ashore for the festivities to return in the early hours half cut. By now only the topmasts of the barges were visible and returning aboard meant climbing down a thirty foot ladder. Immediately outside the barges in mid river were two massive square riggers, TOVARISCH and SAGRES, further down river was the magnificent GOCH FOCH. Some excitment was provided by another square rigger MIRCEA which grounded and start to capsize. Lines from the masts were led ashore and the rescue of her crew prevented a total inversion. The sail Training Association schooners, WINSTON CHURCHILL and MALCOLM MILLER were lying above Tower Bridge and the schooner CHARLOTTE RHODES of television Onedin Line fame was there too.

The following Saturday saw a parade. This was a highly organised event with a strict running order. The instructions required us to be ready to go with our engine running from the early hours. Unfortunately we were scheduled to be the last barge away and just before we cast off the engine decided to die due to a blockage in the cooling system.  By the time it was repaired everyone else had gone. Eventually the engine was running again and we plodded down to Grays. We decided to give the Parade of Sail a miss and anchored off Grays Yacht Club and went ashore for a pint. We were warmly welcomed in the club bar and some us walked into the town. Late in the day we got under way and that evening arrived at Southend. Sunday morning was to be the start of the Round the World Yacht Race but we did not see anything as the whole area was shrouded in mist. We had a very leisurely sail back into the Swale, we passed the ORINOCO lying back at Oare along with C.I.V. In daylight I could not believe that we were actually going to take the barge up what appeared to be a muddy ditch but was actually Faversham Creek but on the top of the tide we swung the barge and put her back alongside the Iron Wharf. Opening time saw us enjoying a farewell pint before ending a charter which no one could forget."

CENTAUR stayed in Faversham for a few more weekends of Club sailing with Fred Woodward again noting thanks to Colin Frake and Jim Gregory for their assistance in piloting her in the creek and she ended her first full season with the Club with a passage on 17-19th October up to St.Katharines Dock, and then into the West India Dock for the winter. The club held a muster on 29th November 1975 when Peter Drew, Chairman of St.Katharines by the Tower Ltd. talked about the economics of owning and running a barge and gave a clear warning that unless about £1,500 was spend in maintenance each year a barge will soon deteriorate. He also argued that reasonable wages should be paid so as to attract young men to become mates and skippers and in order to raise money for these changes, charterers must expect to pay sensible rates and not get their barging on the cheap. His views were not immediately well received by the Club members in the audience but it seems the Committee took them to heart and the wisdom of his words was acted upon. Now, in 1995 some £6/7,000 is spent each year on general maintenance without taking into account sailing expenses, the cost of preservation and major new jobs but the club still does not pay wages.

Bulletin No.277 contained a composition on "Our Barge Holiday" by Elizabeth Eddy aged 8 1/2 which described how our new barge was being used in 1976 on a club weekend under Capt. Leslie Williams, his mate Chippy Wood and Alan Cordell.

"August 30th: ` We got on the barge CENTAUR at half past eight at night. Julie, my cousin and Michael, my brother and I all went below to see where we were sleeping and then we came back up on deck to look around. After that Christine my sister and I went to bed and Julie and Michael came about half an hour later.

August 31st: ` Our first night on the barge was very short. Mummy and Daddy got dressed and went up on deck. Later I helped Mummy to make the breakfast. We had cornflakes and then bacon, scrambled eggs and tomatoes. Then I went back up on deck. It was very cold and wet all that day and we needed out coats on. I had a go on the wheel and soon it was dinner time. Afterward I had a rest. We saw a sunken ship and two lots of forts and we crossed the Shivering Sands. We put the anchor down at Queenborough and Deryck, Michael, Derek and Alan went out in the small boat to look at some barge hulks. When they came back, Julie Christine and I went out. Our meal was ready by then and afterwards we all went to bed.

September 1st: ` We had a good nights sleep. Some of the men went off in the small boat to take pictures and after breakfast we got up the anchor again. It was a lovely sunny day and we could see a long way., We saw three other barges sailing. One was called IRONSIDES. She looked lovely. We also saw where Southend Pier had been burned. We all had dinner on deck but afterwards it got cloudy again. It was time to turn back towards Hoo and the wind was much stronger and coming towards us. Michael and I were in charge of the foresail and

Brian and Julian did the leeboards all the way back. We sailed right into Hoo and had dinner straight away. Afterwards we all packed up our things and cleaned up the barge."

On 25th and 26th October 1976 the club held a further open Day on Greenwich Pier with Geoff. Collard of Cube Charters agreeing to their barges ETHEL and MAY joining CENTAUR and PUDGE. Both our barges went into Junction Basin in the West India and it was agreed that CENTAUR would remain there until it was decided where to refit her stem. Fred Woodward worked on her engine during the winter of 1975/6 and the committee decided that only CENTAUR should be raced in the 1976 season. To get the Club into a more competitive position there should be a specially selected racing crew of five members who should be properly trained, and the Duty Mates should be trained to be skippers. Insurance and other problems were considered and it was also decided to purchase a VHF radio for one of our barges. The Hon. Treasurer was able to repay half the money loaned by members and redeem half the CENTAUR Bonds and Bill Covill took over organising the work on our barges. In 1976 CENTAUR sailed for 23 weekends and of these 7 were at Pin Mill, 1 from Maldon and 12 in the Medway covering some 1750 miles during the season. Vic Wadhams became our main skipper while Terry Everett took CENTAUR when she was in Essex and Suffolk waters but they were only two of the fourteen skippers who sailed our barges that year. The Annual General Meeting of the Club also passed a resolution abandoning the original object of keeping a barge on 'London River' so paving the way to the move away from Greenwich and London Docks to the Blackwater.

Both club barges were included in the 28 craft insured through a scheme arranged by the Club through Alan Peake but after a disastrous 1976 season for the fleet as a whole, Alan was invited to address the skippers on safety at the Medway Match briefing and happily in the 1977 season there was no major damage. Our sailing programme was set at the beginning of the Silver Jubilee year 1977 and CENTAUR could be chartered mid week at the rate of £18 for 5 days with booking again made through Colin Welsby. The Club entered five races, including the Jubilee Thames Sailing Barge and Smack Match organised by the Club in which PUDGE sailed.

CENTAUR spent the winter of 1976/7 in West India Dock locking out on Friday 22nd April under Vic Wadhams with Fred Woodward as mate, Bill Covil, with Peggy Woodward and others as crew. They motored up river to Irongate Wharf to tie up for the night and were joined next day by more members including Doug Nicholls and Valerie Sergent. In a south west 5/6 gusting 8 they cast off before seven in the morning to motor through the Barrier before setting sail. All went well until just after one the topsail sheet parted and they had to start the engine,

71

stow up and drop anchor off Stangate Creek. Fred and Peggy came down the following weekend to help Vic to repair the problems of the previous weekend but the wind was too strong to let them lower down the gear. In the following weeks Vic, Horace Briggs, Gordon Williamson, Terry Everett and Les Williams had the barge and Charlie deCaluwe was duty mate for the Hoo Ness Yacht Club small boat rally on 21st May, but only, Elkenob, reached Stangate in the heavy weather.

CENTAUR did not participate in the matches early in the year but her topsail blew out and was replaced by one loaned by Brian Metherill so that she could follow the Medway Match under power. After her time in the docks and around Kent CENTAUR came back to the Blackwater but by this time her bottom was foul and covered with weed. The skippers decided she was in urgent need of blocking so on Thursday 30th June, 1977, with Barrie Lawrence as mate and W.Bonner, crew, Terry Everett sailed her from Maldon to Pauls Yard at Ipswich. Capt.Mick Lungley whose ship, m.v. RITO locked in that midnight, along with our other skippers Gordon Williamson, Harold Smy and J.Southgate were down at nine next morning to position the barge for blocking after which they all adjourned to Ship Launch Inn to recover and take high tea. She settled on the blocks and they got down in the mud to begin to scrub round. They worked moving the very heavy weed fouling for the next six hours before adjourning to Pin Mill 'for discussion' of the days' work.

Meanwhile Dick and Joyce Wilkins were delighted to receive a postcard to tell them they were booked for '*passage Maldon to Pin Mill, Saturday 2nd July*'. "Hurray!" Joyce recalls "but next day another postcard arrived '*change of programme 'Centaur' Blocks weekend at Pauls Wharf Ipswich*' O well it could be fun! We arrived after dark spied a barge between the warehouses, but it was ENA. They directed us to the gates where Securicor checked whether we wanted the barge or a submarine!. CENTAUR was in darkness and deserted, but there were signed of recent habitation. A whistle and rattle of gangplank announced the arrival of Barrie at a shockingly late hour. He was obviously surprised to see us" The local skippers were all back at eight next morning under the barge to scrape a second layer and a nurse, Nina Zelezinski came aboard with an offer of assistance and was sent underneath while Joyce kept them happy with innumerable cups of tea and coffee and the radio kept them up to date with the latest cricket scores. They all continued with much hard scraping and scrubbing until the incoming tide put a stop to work for the day at ten when they all went to town for a 'nosh up' as Barrie Lawrence put in in the log. They agreed they had found the hull fabric good. Then on Sunday 3rd July they prepared to haul off; Barrie noted the names of the members who turned up and signed off the log 'Where were you all in our time of distress? ?". Terry Everitt and W.Bonner were in charge of the barge which shot away from the shipyard towards Pin Mill in suprise at the loss of months of fouling while Barrie gratefully scrounged a lift with Dick and Joyce back home.

The following weekend, 9th July, with Barrie as mate and a full crew including Cliff Manning, Doug Nicholls, Pat and Peter Boss, Valerie Sargent and Gwyn Lawrence, when Terry Everett took CENTAUR in the Pin Mill Match. The wind was a fresh N.N.E. giving a lively days sailing to the 20 barges which took part over a 20 mile course to the Medusa Buoy. Classes B and C were timed to start at 9.30 and SPINAWAY C closely pressed by IRONSIDES luffed hard up to CENTAUR who had to sheet in hard to luff up to the Levington shore, but all cleared and we were on course; just, with our barge going out in front. With Cliff Foot buoy to port, Barrie had just time to note in the log "just astern of VIGILANT in Class B, the barge going well and being sailed well, ETHEL ADA three lengths astern". CENTAUR crossed the line at 14.50 and brought up off Potter Point. Barrie was pleased to note what we were first in class C at the outer mark and at the finish beating FELIX into second place. Sunday was spent on board in the various tasks which always seemed to be needed to be done when Barrie was about and he wrote in the log "Our thanks to Terry for a smashing sail and for organising the blocking the previous week-end all very worthwhile and necessary. Also thanks to Jack Haste and Committee of P.M.S.C. for a grand 'do;'

Weekend sailing continued as usual with Terry again skipper the following weekend when Charlie de Caluwe took a full crew into the Walton Backwaters to moor on BRITISH KING's buoy. He had only just squared off back at Pin Mill and was about to depart when members of the Society for Spritsail Barge Research came down the hard for their five day charter arranged by John Eddy with Les Williams as skipper and his friend 'Chippy' Wood as mate. Then followed weekends with Pat Fisher and Cliff Manning as mate, Gordon Williamson with Fred Woodward, Rupert Neivergelt, then with Barrie Lawrence and Pat and Peter Boss.

CENTAUR's hull scraping paid off in the Southend Match with the wind ESE force 2 the start was delayed by an hour and FELIX was first over the line. CENTAUR set off on a run to the W.Leigh Middle when, while taking in a boom to reset the sail on the stem head the mate noticed a ten foot split in the topmast from about ten foot above the top cap to the heel. It seems that a scarf in her topmast, which had been made some time earlier, had sprung. The opening of the split prevented the crew lowering the topsail as the hoops got caught in the split and they had to send Tony de Winton aloft to guide the hoops over the split to lower the topsail without damage. Despite the split, Skipper Gordon Williamson decided to continue racing under jury rig and they quickly rigged an extension wire to the heel rope and housed the topmast to bring the split between the caps. They lashed the heel to the mainmast with a preventer rigged on shortened topmast stay, and stopper preventers were rigged on standby backstays using the guys from the setting boom. Under the skippers' directions they removed tackles for runners and led the wire through the snatch on the rail. Tackles led horizontally along the main shrouds were hove taught to set up the topmast in its new position and they checked to see that the spit was housed safely at the mast head. The headstick clewed home and hauled out, the topsail sheet re-set, the lowered staysail sheets led direct to the rail aft, set all square and she continued on the race. On course to the Mid Shoebury Buoy; they rounded and returned to the pier head in a severe thunderstorm with torrential rain. Mr. Lawrence noted in the log 'the barge very clean', but they received instructions from the Officer of the Day to shorten course and return to the start line off the Pier for the fun to finish at five oclock. Hoo bloody ray! a dripping Mr.Lawrence noted. They anchored above the Pier; all squared away and stowed up by twenty past. All below to dry out and prepare to be picked up for the match supper where chatter about the sailing was somewhat subdued by the sad news of the death of Claud Curtis the chairman of the Match Committee.

At a meeting of the Club Committee on 6th September 1977 the replacement of the sails on CENTAUR were discussed with estimates from Jim Lawrence on the table, of £2,027 for replacement in flax or £1,750 in duradon. The cost of repairing the stem was estimated to be £1,650 and a new topmast would cost about £1,000 in addition to which new anchor chain was required. The expenses were far beyond those to which the committee members had been accustomed and they seriously debated disposing of CENTAUR. The difficulty of finding a replacement barge would pose even greater difficulties and a decision was deferred.

Meanwhile the duty mate for the weekend 18th/19th September was again, Charlie de Caluwe. He had decided to retire from the job of representing the Club on these weekend sailings and swallow the anchor as he was moving to Norfolk. The list of the crew he had received from the Crewing Secretary did not have any names he could recognise. A couple of new members came aboard and no doubt as was his custom, he showed them where to stow their gear, then took them on deck in the fading light to pull the ropes and tell them what he expected they might be called upon to do when they set sail. He must have been somewhat worried that no-one else showed up as the tide began to flood and the time for departure neared. But then Horace Briggs, the skipper and a past crewing secretary and duty mate appeared, followed by Commodore Peter Love, and his wife Valerie, past crewing secretary, Jack Stocks, and Tom Johnson both Past Rear Commodores; Gordon Chell and his wife; a past Rear Commodore and Bob Wood, a Vice President, Lew Foweraker past Vice and Rear Commodore and editor of the newsletter and to complete the list past Commodore Barrie Lawrence. After sailing to Queenborough (via Harty Ferry where they sent a boat ashore for horseradish (where they thought they would get it was not disclosed) Charlie wrote in the log "After nineteen years as a duty mate this in all probability is my last Hoo to Hoo weekend. I am delighted that so many of my old friends honoured me by sailing in this crew".

At the October Committee meeting a decision to replace the stem was essential, if the work was to be done that winter, and not for the first time Barrie Lawrence came forward, thumped the table and got an agreement to find and spend the money. During the past three years the Club had earned sufficient money to pay off the rest of the purchase price, plus enough for the new stem and apron. It was decided that with the 1978 subscriptions we could afford the new mainsail, topsail and foresail to replace those which were now getting to the stage of patching the patches. These new sails were ordered from Jim Lawrence of Brightlingsea at a written quotation of £2,572 to be delivered in time for the 1978 season and arrangements made with Fred Webb to take CENTAUR to Pin Mill for work to her hull.

The new suit of sails, except the patched mizzen!

6. Restoration project
1977 - 1981

Pin Mill is a long way from Kent and the various London docks which the Club had used for winter maintenance since it had been formed and a lot of the regular workers could not travel the distance to Suffolk at weekends, especially when PUDGE was moored much closer to home. Peter and Pat Boss had been Club members for some years but they had spent a lot of sailing time in barges outside the Club, particularly on VIGILANT. They had long associations with Pin Mill and many friends there and were quite happy to get involved. Barrie Lawrence, of course had many friends there including, Richard Duke the previous owner with whom he remained in contact so gradually a new working team evolved which trundled up most weekends.

CENTAUR was moored alongside the MAID OF CONNAUGHT part way down Pin Mill Hard and you can imagine the scene on arrival around 8.00 p.m. Friday evenings, pitch dark and freezing cold. It wasn't too bad when the tide was out. All the working party had to do was to don wellies, walk down the Hard and scramble up the ladder left dangling over the side. When the tide was in it was a different matter. First they had to get the barge boat which had been carefully moored in a position judged to be accessible according to their prediction of the position of the tide, made the weekend before (They were not always right!). This led to some interesting experiences, most of them very wet and cold, and why is it that they always make wellies and waders an inch too short?

On one occasion the tide was out and CENTAUR sat squarely on the bottom. The working party got on board without mishap and Peter nipped down the companionway stairs in the dark and stepped straight into a foot of water! That's another strange thing, water never seems top have any problem flowing in but it finds it difficult to drain out. Fortunately, the battery rack was still above water and the team were able to start the engine, jury rigged to take the cooling water from the bilge and pump it overside. Two men were stationed to pump water through the loo by loading it from buckets and yet another operating the hand bilge pump. In two hours they had the water level down to the bottom of the saloon stove and lit all the fires on board retiring to the Butt & Oyster. On their return they had a dry although very damp barge (if that's not a contradiction in terms), but they had enough of Pat Watt's bitter inside them to be sufficiently anaesthetised not to notice.

Fred Webb removed the stemband, unbolted the old stem and cut out the apron. A new oak apron was fitted and the original planking spiked back into position. The new stem was then fitted to the apron and, most importantly, a new iron stemband was forged and bolted through. The stemband is very important because the whole weight of the gear hangs on the forged eye at the stemhead. Fred had to drill through the forefoot and stem to fix the bottom end of the stemband and found the floor frames forward were over twelve inches thick. The length of the hole was 42" bearing out the legend of the strength of her construction.

The next job was to bend on the new suit of sails. The Club had not rigged a barge with new sails before so it was with a little apprehensive but with a long list of do's and don'ts from Jim Lawrence the working party managed to do it right. However, Peter was sure that Jim had got the measurements wrong, everything was so small! Since then we have seen several new mainsails bent on but it never ceases to amaze how undersize they start off and how quickly they grow to fit.

CENTAUR was completed and rigged ready for a shakedown cruise to Mistley on 1st May to stretch her new sails, Then blocking on 6th May 1978 to fix the stern to be ready to sail just one weekend before the Medway Match. She came off the Hard on the Friday night bound for Hoo but within an hour she was leaking so badly that the trip had to be abandoned. The leak was not from the stem as might have been expected, but from the port side aft, the crew could see where the water was coming in from the inside, but not where it started off outside. Barge hulls are double skinned and are caulked between the skins. A leak can start outside at one point, run along between the skins and appear many feet away. She was beached again and Saturday was spent making temporary repairs without success. Finally on the Sunday morning at 5 a.m.. as the tide came around the barge, Terry Everett then Skipper lay on his back in the engine room with his head stuck through a hole in the bulkhead into the locker aft where the water came through a joint in the planking. Peter donned his thigh boots and went over the side with a team of messengers ranged between. The wash from the ships going by sent Peter scurrying for the ladder but after he discussed once again that his boots were an inch too short he gave up. Terry suddenly called out ' It's pouring in". The message was passed out to Peter and he marked the spot where the water had reached the joint in the planking. The fault turned out to be a rotten section of planking under a piece of doubling on the runs aft. It really needed the planks to be replaced as it had already been botched once.

But there was no time and no money so she was left on the hard for Fred Webb to patch her again. The next weekend we headed for the Medway to show off our new stem and sails but it was touch and go whether we would get there in time for the Match. We did, just, by Terry sailing her straight onto the starting line.

At the end of the 1978 season we returned to the London Docks but apart from normal maintenance all we did was to cut out the old steel water tank put in a new glass fibre one and fit a new hatchcloth. We were saving up hoping to get enough money to start the next big job, the restoration of the stern at the end of the 1979 season. At the April Committee meeting Peter Boss proposed that no further sale of CENTAUR Bonds should take place since she had now been fully paid for and was now a viable proposition. While her starter motor was still giving problems in June 1979 she completed 8 weeks sailing out of Ipswich in the busy Harwich estuary . The Commodore had to tell the Committee of a complaint from the Harwich Harbour authorities of problems contacting the barge while she was in their waters. There was no record of any particular difficulties in the log, but in those days very few barges had radio and we were by no means alone in sailing without one. However, the committee agreed to VHF radio being bought for both barges and the Club agreed to pay for teaching the duty mates.

In the late summer of 1979 the engine seized. An examination showed that we needed a new crankshaft which meant a complete stripdown, so we had to finish the season without an engine. The role call of skippers that year included Gordon Wilkinson, Harold Smy, Mick Lungley, Tom Polly, Terry Everitt, Stan Yeates, Horace Briggs and our two members David Ward and Peter Boss who, during the previous year who had been encouraged by Vic Wadhams to take on this responsibility. During the winter Bob Pink who was a marine engineer, with one or two labourers rebuilt the engine while she laid at Hoo. Although Bob gave his services free, the materials and parts were expensive so we blew some of the stem money. We also had to spend £4,500 on replacing sections of the inwale and deck carlings on PUDGE so we were broke once more.

1980 was quite a good season. That means we didn't break anything and all our earnings went into the bank. Val Sargent found an anchor at Woolwich which was changed over for the old one during the May Bank Holiday in half an hour. On 16th August, 1980, Stan Yeates was sailing CENTAUR once again on one of the Hoo to Hoo club weekends. Terry O'Sullivan writes "The tides were around half four to five o'clock and a trip round the island was proposed. We were then normally moored on the Upnor side of Hoo and under Vic Wadhams we would normally sail outside the Hoo Island to get down river. Stan was still occasionally sailing SPINAWAY C and often used the winding inside channel and this was what we did that trip, a new experience for many club members.

Once into the main channel and we set the sails with sw-w wind 3 - 2 and made a pleasant passage outside Sheppey. The wind fell a bit lighter but still enough to sail and after passing the Spile Buoy we would see another barge in the Swale mouth. We were moving slowly over the ground and close to it, as the leeboard warned us a couple of times. I remember having a long spell on the wheel because Stan liked to play with the sails and he also brought out the staysail. The other barge turned out to the SCOTSMAN out for a trial sail with Colin Frake. She was rigged except for the mizzen. As we passed close by we exchanged pleasantries and enquired what was up the creek. Colin replied that MIROSA was away as well as themselves so that if we wanted to go up there would be room.

Well the tide was right and the opportunity doesn't arise that often so after beating in past the Shell Ness and the Horse we turned up the creek with all sail set. As we approached Hollowshore the wind was abeam and gave us a fine sail. Laurie Testers' tug came away from the yard and passed us on her way out. Aboard was Laurie himself who shouted to Stan "Who gave you a fair wind up my Creek?" Stan had us turning up till just past the dreaded pylons where we stowed up and motored up the rest of the way. As we approached the Iron Wharf Alan Reekie greeted us and said to swing there and lie in MIROSA's berth A pleasant evening was spent in Faversham with a few drinks in the Anchor and a brief chat with Alan Reekie who was having a barbeque on the wharf.

Stan Yeates

Next morning we were underway just before five out of the creek under motor, We gave a start to a couple of yachts anchored in the channel just below Hollowshore leaving little room as they hadn't expected a barge to pass by that morning. The wind was still the same direction as the previous day, but a little stronger giving a brisk sail back to Hoo. Stan's only comment about his old charge that she was a bit hard headed rigged as now and was more balanced when she had a bowsprit; however we all enjoyed our sail that weekend finishing off by sailing alongside the lighters under topsail, with all lines secured by just after four.

The West India Docks closed in July 1980 and no alternative berth was offered by P.L.A. A winter berth was, however, found for PUDGE in the Pontoon Dock in the Royal Victoria system for the winter 1980/1 and members had to contact John Kelland to work on her while CENTAUR went to Pin Mill again. Peter Boss again takes up the narrative "By August we had decided that if we included the money from the 1981 subscriptions we would have enough to carry out the work needed to the stern. We knew she needed a new transom, part of it had completely fallen away and was covered with a plywood patch. One section was so badly split that the skipper could lay in the port bunk in his cabin and see where we had come from. There was rot in the sternpost, the fashion frames were almost non-existent and there were soft spots in the wale and inwale. We decided to go ahead and kept our fingers crossed that was hidden was better that we could see. Naive you think! Babes in arms would describe us better".

After checking around we had finally agreed that the job should go to Joe Dunnett of Pin Mill We had seen Joe's work on the barge ETHEL ADA which he was restoring for Geoff. Mellor, also the work on his own fishing smack DOROTHY which he had completely rebuilt and we were very impressed. We arranged a deal with Joe and his partner Mike Newman that we would pay them for the hours worked and would buy the materials. We would also do as much of the ripping out and general labouring as possible.   This arrangement suited us very well and would help keep our costs to the minimum. CENTAUR was taken to Pin Mill at the end of October 1980 and laid stern first on the beach between the hulked barge RELIANCE and the housebarge LESLIE WEST in the berth where the CHARLES HUTSON had lain for many years. That put CENTAUR right outside the Butt & Oyster car park, just 25 yards from the pub door. Could you wish for a better place to work?

As with the stem job we had a small, but keen working party there for most weekends only this time it was a doddle getting aboard and ashore. CENTAUR was so far up the beach that she only lifted on high springs. We could even get ashore for a lunch time pint at high water. Joe built a set of stairs to deck level and with a plank to a trestle and another to the top of the car park wall you could be in the Butt in three bounds. No prizes for guessing who worked that one out, particularly when Mike Collins was one of the team.

Our first job was to take out the interior of the skippers' cabin. CENTAUR has a fine example of an original traditional mahogany panel cabin which needed to be preserved and Peter Boss who had taken charge of operations was relieved when Rob Dudley volunteered to dismantle, restore the panels and rebuild the cabin. So the cabin interior was removed piece by piece. The steering gear was then dismantled, the rudder lifted off and the stern post and transom cut out. At the same time the saddle chock fell apart and the plank ends of the deck, wale and inwale disintegrated. It was time to start rebuilding before the watchers from the Butt & Oyster thought we were breaking her up and came for the firewood.

The first batch of oak arrived and Joe and Mick shaped up the planks for the transom and carved the name 'CENTAUR' and the Port of registration, 'HARWICH" in their own style of lettering. The transom was then built into the barge, the fashion pieces fixed inside and the new sternpost bolted on complete with newly forged gudgeon irons. The next job was to decide on the condition of the cant frames. We removed a section of the inwale and were then able to examine the frames properly. Not a pretty sight. There was no going back and no point botching. Replace the lot, both sides to the mainhorse!

The rails and covering board were cut out, then the rest of the inwale, all the linings and the outer wale. Some of it was very difficult to remove while some came out in handfuls. The shipwrights took patterns of the cant frames and gradually replaced them one by one right through to a point level with the saloon bulkhead which is approximately two feet forward of the main hatch after headledge. To do this we had to remove everything in the way. This included sections of the skipper's cabin and galley bulkheads, all the engine room fixtures and fittings; engine propshaft,

exhaust and cooling pipes fuel & water tanks, electrical controls, batteries etc. The 12 volt system was disconnected and we rigged up temporary mains lighting for the working areas and reverted to tilley and oil lamps elsewhere. Work progressed through the February: gales blew through the non-existent sides and areas of missing bulkheads straight into the saloon. Old blankets helped to keep out the draught, but not the cat from the LESLIE WEST who crept in at night to steal our breakfast bacon. This led to one or two interesting debates between the cat lovers and the not so cat lovers on how we should cure the problems. The cat lovers won and we locked up the bacon, a pity thought Peter as he had some ingenious ideas.

As soon as the frames were in the shipwrights fixed the outer planking along both sides. They varied the lengths of the planks, staggering the joints over ten feet or so to avoid ending in a straight line. They then fixed the new wales from the transom to the leeboard sheeve, chocks and inwales to match. It was now time to turn attention to deck level. Both the beam supporting the after end of the cabin top and the headledge had areas of rot in them, as did the side deck carlings and the lodging knees. This whole area had been jacked up and supported by props while the sides of the barge were being removed. There was no point in spoiling the ship for a ha'p'orth of tar (or another £1,000 worth of oak) so out they all came, and in went a new beam and headledge plus new carlings and knees from the transom to the mainhorse. New lining was then fitted as far as the galley and we were finally able to replace the galley bulkhead, although it was no longer necessary to keep out the winter chills. Winter had long gone, it was May. The sailing season was well underway and we still had a barge in bits. So much for our carefully timed programme.

To be fair everyone had carried out much more work than originally intended and the wettest March on record had not helped either. We decided that if we did not call a halt to the work we would be on Pin Mill beach all summer. To complete the stern area the cabin top, forward headledge and decking between the cabin and the main hatch really needed replacing along with the beam through the engine room. However that would now have to wait for another day. Even so the shipwrights still had to make and fit the saddlechock, covering boards and rails as far as the mainhorse, lay and caulk the side decks, rehang the rudder and install the crab winches, bollards and davits. To keep us occupied while all that was happening the working party re-assembled the steering, put the engine room and workshop back together, refitted all the engine controls, cooling water and intake pipes, exhaust and pumps. Then we put back the fuel tank, but the electrics were in such a mess that Doug Nicholls decided to rewire the whole barge and make a new control panel for the engine and battery charging system.

There was not time for Rob Dudley to fully replace the skippers cabin but we had already restored the panels of the pilot berths, so he installed those and made the bunks so that the skipper would have somewhere to sleep. The rest would be done later when time permitted. It was now the end of June and we were getting desperate. We had already missed the Medway, and Blackwater Matches but following a weekend off to get afloat on other barges for the Pin Mill Match, Peter

Boss and the working party put CENTAUR on the blocks at the end of Pin Mill hard. They scraped her bottom while the shipwrights refitted the propeller shaft 'A' bracket into its new position on the no longer sagging stem. The engine was then realigned, connected to the prop shaft, the rigging set up and we were finally completed and Peter Boss sailed her off the hard in the third weekend in July.

The 1981 sailing season ensued. Alan Cordell had retired as editor of the Bulletin to be succeeded by Elizabeth Wood and she continued to publish reports on CENTAUR from Peter Boss. His wife, Pat had a display of photographs of the winters work on board which gave members some idea of where we spent all the money. CENTAUR remained in Suffolk for five weeks before making the passage to Kent for the Swale match on 22nd August in which she was placed 4th in her class and both barges finished the season sailing out of Hoo. On 15th October both barges came 'up river' for the now traditional Open Day at Greenwich and were joined by LADY DAPHNE under Derek Ling on loan from Taylor Woodrow and later XYLONITE came alongside for the night. Sunday made up for a distinctly wet Saturday and had it not been for reviving cups of tea and coffee the stewards might well have suffered from hypothermia, but despite the rain more funds were put in the coffers for our next project, the restoration of the bow section.

Joe Dunnett was prepared to take on the work on the same basis as the stern job, The berth at Pin Mill was still available, the only problem was again the money. We were sure that the bows were not as bad as the stern and would not cost anywhere near as much. We had one or two ideas of raising monthly if we did find ourselves a little short so we decided to go ahead. In Mid October 1981 CENTAUR went to Pin Mill and was put in the same berth as the previous year. A working party lowered the gear removed the sails which were left in store with Jim Lawrence. The forestay, stayfall and windlass were removed so that the foredeck was clear for the shipwrights to start work. First they stripped off the port rail covering board, outer wale and three planks from the stem to the rigging chock so that we could see the condition of the frames. We expected to see some rot and were not disappointed. There was some rot in most of the frames. She must have had a major rebuild job sometime in the past as there were quite a number of double frames in the forward part. We were now in the same quandary as with the stern, effect a repair or completely restore. If we had known just how bad the situation was we would have waited until we had more money to spend, but it was now too late. We had to remind ourselves that we were not just a sailing club but also a preservation society and as such have a responsibility to preserve our barges for the next generation. It had to be a complete restoration, whatever the cost. We had to find the money someway, beg or borrow it. So we said to Joe "renew everything that isn't sound" just as we did with the stern.

We stripped off the rails, covering board, lining, inwale, outerwale and planking for approximately 30 foot along the port side. We jacked up the deck beams and renewed the oak frames one by one which came to a total of seventeen all placed at 19" centres. While this was going on one team was working on the general winter maintenance, overhauling the standing and running rigging while the rest of the volunteers stripped out all the accommodation that was in the way of the shipwrights. Out came the foc'sle bulkhead, chain locker floor, bunks and lockers. Then six of the eight bunks in the two port cabins were dismantled, the only ones remaining being the fore and aft bunks built inboard at the side of the keelson. During her trading days CENTAUR had had her hatch coaming sheathed in steel to stop the unloading crane grabs from chewing them up. We stripped off the sheathing from the forehatch to make sure the coamings were not rotten and could take all the fastenings for the new deck carlings. Fortunately they were still good, so that was one thing at least that we did not have to renew. As the shipwrights progressed aft along the portside we found it necessary to strip out the lamp room door, bulkhead and shelves as the new framing was to right up to the rigging chock which is above the lamp room. Where the frames had been previously doubled, the newer frames were still in good condition and we decided to leave them in but replace all the old ones thereby keeping the extra strength with the double framing.

It was now December. We had only just started the job and we were already in trouble for money. We had been banking on getting in sufficient charter deposits to tide us over to January when we start to receive membership subscriptions. However, bookings were slow to come in and the restoration work had progressed faster than anticipated. All the wood for the planking had arrived, along with a bill for £2,700. We owed £950 for oak frames and £1,500 for labour. With incidentals like bolts and spikes to add we needed £5,800. We had a temporary cash flow problem but three members made generous loans to tide us over till funds rolled in. It was now time to start ripping out the starboard side. We took out the bunks in the forward cabin and the washbasin unit from the washroom. This left us with only one four berth cabin and the skippers cabin intact which gave sleeping accommodation for only six people. However, all the running rigging had been stored in the aft starboard cabin, so the only place left was the saloon. Each night when the evening meal was over and before we all evacuated to the Butt & Oyster, everyone would lay out their sleeping bags, There were bodies on the dining table, the galley counter, the narrow keelson seat and all over the ceiling. The side deck leaked, so unless you were sure it would not rain in the night it was fatal to sleep on the side locker seats. With everyone dossed down the saloon looked more like Fagin's kitchen! The only compensation was with the stove well stoked up it was warm. We never got to the stage of having more people that were places available as only the hard and dedicated (or perhaps just the lunatics) stayed overnight.

The frame replacement continued on the starboard side until the whole section from the stem to the rigging chock was completed. By this time sitting in the heads had become a little 'adventurous'. The outer planking and the inner lining had been removed, so any anyone using the loo could have a good view of the Butt & Oyster and of course the patrons of the pub could take the opposite view! A length of old curtain hung up helped to give a little privacy, but didn't do much to stop the cold northerly wind whistling round the stern. However, this condition didn't last for very long. The shipwrights removed the heads' compartment floor so we didn't have a loo at all. This seemed a convenient moment to overhaul the pump unit so Peter took the whole lot home to give him some occupational therapy during the long dark winter evenings. There is a public toilet at the top of Pin Mill hard, only 100 yards from the barge and during opening times we could use the loo in the Butt

& Oyster. But Sunday morning was a real test of endurance. The public loo was locked at night so around 8.50 on Sunday mornings one would find a queue of bargees standing outside in the freezing cold with pinched cheeks, waiting for the man to arrive with the key at nine. It was a good thing he was a very punctual man, as I dread to think of the consequences of his being late.

In the next phase the shipwrights planked up both sides, bent and fitted on the outer and inner wale and spiked on the lining. They then fitted the five foot high 4" x 2" iron knees which were bolted through the floor and side frames, two each side, along with the two big iron breast hooks, all of which had been chipped and anti-rusted by our crew. By this time we were into February and we had run out of money again! We had well under-estimated the size of the job, and just how much it would cost: there was no going back.

The new foredeck structure , & the pub!

We has spent over £9,000 in labour and materials and we were only halfway there. We still had to buy the wood for the new windlass bitts and their knees, two deck beams, deck carlings, deck planks, rails and covering board plus fo'c'sle bulkhead, chain locker and floor. We discussed borrowing from the bank, but we really could not afford the interest.   A number of long standing committee members once again rallied round and loaned the Club a further £3,000 which got us through the next month while we made another appeal for help. The Society for Spritsail Research immediately came up with a £1,000 loan and Maurice Butters also answered the call for help. Over the next month a further 11 members made loans amounting to £900, and we got more Life Members which brought in a further £500.

We were rich again and there was no stopping us. The next job was to rip off the rest of the deck from the rigging cocks to the stem and take out the windlass bitts and their knees. Aft of the fo'c'sle bulkhead the deck carlings, the short beams   between the barge sides and the hatch coaming supporting the side decks were made of iron and all had rusted badly with half of inch of scale built up on top.   This had pushed the deck up and drawn out the coach screws so that the deck was floating and not held rigidly to the carlings. We decided to replace all the iron carlings with wooden ones. The iron beams supporting the fore and aft ends of the forehatch were still in very good condition, apart from the same rust scaling effect as the carlings so the forehatch was jacked up while we chipped off the scale and rust-protected them.   The winter had not been too bad up until then, but as soon as the deck was removed it snowed, long and heavy. One of the first jobs to do each Saturday morning was to dig out all the wood shaving dropped down into the bilge during the previous weeks work. You would not believe how many dustbins full had to be humped up a ladder and heaved over the side. It was bad enough with dry shavings, but when mixed with bilge crud, soggy and covered with snow, the job was not only dirty and nasty but back breaking as well. Whenever Brian Dawson and Martin Phillips talk about the bow restoration winter they always refer to doing that particular job and say, "How is it we always drew the short straw?" Well, they say that hard, nasty jobs like crud shovelling are character building, the sort of stuff Commodores and Club Secretaries are made of. It must be right, they both made it.

Next the shipwrights scarfed new ends onto the ninteen foot long wooden beam al the start of the round of thc bow. They sited a new ten foot long beam through the middle of the fo'c'sle and this was followed by new windlass bitts with their thrust beam. Lodging knees were fitted at the ends of all the deck beams and wooden carlings along the length of the forehatch. This completed the framework for the foredeck and enabled them to lay, caulk and pay new decking and covering boards from the stem to the forehorse and then to fashion and fit the two large knees that take the thrust of the windlass. The barge was now weathered in again.

The shipwrights next job was to fit the rails. These were to be laminated from three thicknesses of Iroko, so they had to make up formers to bend them to. Before they got started on the rails we asked them to complete the work down below so that we could make a start on putting back the accommodation. They built the fo'c'sle bulkhead and chain locker and fitted new raised floors in the washroom, heads and fo'c'sle and then returned to continue work topsides. Mike Collins took charge of the cabin and lamp room rebuilding team, refixing bulkheads, bunks, shelves coaming covers and electric lights. Pat Boss with Pat Collins and occasional helpers cleaned and painted or varnished just about everything in sight. The mates discussed how their new fo'c'sle should be arranged where and how many bunks, double or single, what lockers, shelves etc. After agreement Rob Dudley volunteered to do the job so he got on with that while Peter Boss tackled the washroom and loo. For a long time we had talked about the inconvenience of having to cart buckets of water to flush the loo when the barge was dried out. So, Peter designed a system where the flush pump operated via two sea water tanks housed under the bunks in the forward starboard cabin. These would continuously fill by gravity while the barge was afloat and keep twenty gallons of water in reserve for when she dried out. This was installed along with the newly rebuilt loo unit with all new chrome plated fittings, with its original but now highly polished, wooden seat and lid (none of your plastic rubbish for us). The installation was completed with a suitable ceremony with all hands present and conducted by our then Commodore, Barrie Lawrence, of which we had better make no further comment. The ceremony that is, not Barrie and particularly the comments on the use or misuse of such a beautiful seat.

By that time Rob had finished the fo'c'sle making one single and one double bunk (in case the Mate took a mate), with lockers and shelves etc. The accommodation was all replaced and decorated and the shipwrights had fashioned and fitted the rails, bollards and chain plates and then carved and fitted the bow badges. During this time was had re-assembled the windlass and painted all the new woodwork with primer, undercoat and topcoat. We picked up the sails from Jim Lawrence at Brightlingsea and rigged out.

Having completed the rest of the deck painting the last job was to put her back on the blocks to scrape the bottom and paint the sides. This was done one cold weekend in April. The blocks at Pin Mill are at the lower end of the hard, not far above the low water mark, and about twenty foot from the up-river edge. The hard had a concrete strip running down the up-river side with a straight drop down about two foot into a stream called "The Grindle", which runs along the whole length of the hard keeping the bottom scoured and making it possible to land from a dinghy alongside the hard and above the tide line. The blocks are quite low and it is very muddy underneath. One virtually has to lay on one's back on old hatch boards kept specially for the occasion to scrape off the barnacles. We had completed the bottom and just had to finish scraping round the chine; We were running out of time as the tide was coming in fast. Peter was making his usual remarks to spur people on to finish before the water got too deep. As we completed the last bit with the water lapping the tops of peoples wellies everyone made for the ladder. Walking backwards with scraper in one hand and dragging a board Peter shouted out "All we have to do is to clean the tools and the boards before we climb back on board". He stepped backwards straight off the edge into the Grindle and disappeared under the freezing water. When he surfaced and dragged himself to his knees those watching said he looked like King Neptune rising out of the waves. Still clasping his scraper like a trident Peter expected some help but all around looked like a scene from a Giles cartoon with everyone falling about in hysterics.

The following weekend as we sailed off Pin Mill hard Peter reflected on the past six months. We had taken on quite a project to be completed one winter, especially for a bunch of part timers, excluding the shipwrights of course. They had done an excellent job worked very hard and put in some long hours to get finished, With the bow restoration complete we had passed another milestone in the history of our barge but even then Peter was hatching up plans for work in winters to come.

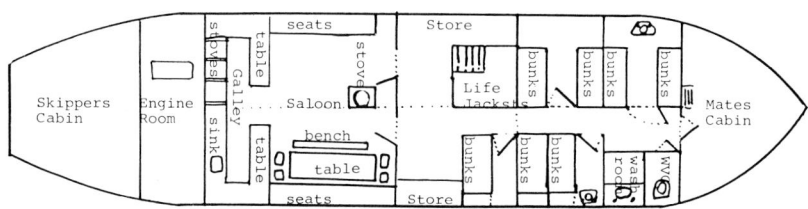

Accomodation Plan that evolved during the next decade
(from Club Charter Brochure)

7. A decade of Club sailing 1982 - 1991

But before then the Club had the 1982 sailing season. This started with gale force winds which restricted sailing but on 15th May a full crew brought CENTAUR over to Hoo for the Small Boat Rally in Stangate Creek and the Medway Match. Peter was forced to retire on drifting too close to the yacht moorings off Gillingham at the start in a close contest with WILL and ARDWINA.

Peter had arranged for Gerard Swift to skipper CENTAUR in the Pin Mill Match on Saturday 26th June, 1982 and to avoid leaving Ipswich Dock in the early hours of Saturday morning it was agreed that Terry O'Sullivan as mate and a small crew including Doug Nicholls should move the barge down to Pin Mill on the Friday evening. Terry O'Sullivan reported that " We met in Ipswich on the Friday lunch time and Gerard broke the sad news that we had lost ever faithfull Vic Wadhams who had been in ill health since the winter and who died at home early that morning. As we left Ipswich at three that afternoon we flew our Red Ensign as a sign of respect for Vic. I thought that we would motor down to the Clamp as the wind was against us and was only a short trip but not so, Gerard had us set the sails before we got to Woolverstone and there we were tacking down river. The moorings are very restrictive off Woolverstone and as soon as we were under way we were about again and a short but hectic sail ensued. By 16.30 we stowed up and motored from the Clamp just inside the mooring by Fox's Bottom.

The rest of the crew joined us on the Friday evening and after a few adjustments to the gear tuning it to Gerard's satisfaction we attended the Briefing and enjoyed a few pints in the Butt & Oyster. The race saw winds SW., 5 - 6 gusting 8 and although we started with all sails including staysail at 7.16 this was soon dispensed with. Unfortunately despite Gerard's best efforts it was not to be our day and we managed to catch a lobster pot just outside the harbour which almost stopped us dead. ARDWINA, OAK and WILL gained places but we still we had a very pleasant day with no damage. The course took us out the Medusa and back, crossing the finish line at 13.45 but we were first on Pin Mill Hard by 14.30. Ashore in the evening for the presentations we all enjoyed the usual grand buffet provided by the Pin Mill Sailing Club. Having a break off the CABBY was my friend and Club mate Cliff Manning. Gerard said we were, unusually, going to have a sail on the

Sunday, so as Maurice Butters had left, Cliff came back aboard with us. So instead of the usual quiet day on the hard we were up and away at 5.30 next morning motoring down to Clamp House for breakfast, setting all sail for a trip down to Harwich Harbour, then up to Parkstone Quay and back up the Orwell. We stopped off at Woolverstone at 15.14 for fuel, then back up to Ipswich Dock to moor up by 16.45." Sadly Cliff was drowned later in the year on the evening of 1st September in a punting accident on the river near his home in Cambridge.

CENTAUR then had a series of charters to Donald Pinn, Colin Welsby, The Thames Sailing Club and then the 1st Dartford Campbell Venture Scout Unit with Mick Lungley as skipper and the charterer, John Kelland as mate. They sailed into the Walton Backwaters and entertained on board the crew of an R.N.L.I. inshore rescue boat and drank and made merry slept, sunbathed, and walked the foreshore after rowing ashore and, of course sampled the local ales. A party from the DAWN were in the Jolly Sailor at Orford and both crews had a musical evening led by Mick Lungley. On Tuesday they sailed on to Maldon where the dynamo belt failed but they tied alongside BEATRICE MAUD while taking on water and victuals and a replacement belt was fitted. Next day while making their way past Clacton in variable force 3 - 4 wind the topmast gave way. It had been repaired after the 1977 Southend Match, and the old scarf joint about six foot above the top of the mainmast just rolled apart. Fortunately the bottom rings of the topsail were still on the stump of the topmast and stopped the mast and sail going overboard. The starboard topmast stay pulled out of the rigging chock, but the weight was taken by the deck fitting and no other gear was damaged and nobody injured. They returned to Brightlingsea to report the damage to the Club Secretary.

Next day they set off but three hours out the forward steering nut failed, again they were off Clacton! It was a welding failure which pushed the steering shaft support block away from the top of the skippers cuddy. There was a large coil of 2 inch rope aboard and the skipper jury rigged the rudder to continue on the way to Pin Mill. John reported that an agile crew member held a colleague, "Tiny" over the stern andTiny managed to thread the rope through the kicking strap chain. Using both leeboard winches, each with its own crew, the skipper was able to resume steerage. An hour later the mate noticed that the engine cooling water had lessened, the belt had stretched and the engine room had water above the ceiling. The electric bilge pumps were switched on and he found that it was necessary to keep the engine jabsco belt tight by placing a weight on the belt. The bilge inlet was continually being covered by wood debris and he decided to stay in the engine room to keep the pumps working. Of Walton Pier the skipper called down that the jury rig ropes were fraying and they were not covering any ground, if anything they were going backwards, so the sails were taken off and they anchored. The weather had deteriorated, a force 6 - 7 being forecast so the skipper and mate decided to call Harwich to see if any assistance was available. Listening in were the Walton Lifeboat who, by chance had a BBC TV crew with them recording "A day in the Life of Walton-on-the Naze". Although the barge was in no immediate danger the Media spung into action, the lifeboat was launched and within half an hour a towrope was aboard. There was then a problem in getting the anchor up in the heavy swell and the dog failed. It turned out that the Coxswain knew Mick and was only too pleased to tow the barge back to Pin Mill.

John continues the story "There was still fun to come; after being towed for perhaps an hour or so a large coaster was seen heading for us with the apparent intent of passing between us and the lifeboat! For what seemed a lifetime both the lifeboat and the barge skipper using their radio's tried to get the attention of the errant vessel. I cannot recall if any reply was received but a great deal of relief was felt when only a few yards away she changed course. So we finally got to to Buttermans Bay and dropped anchor but due to the weather being so bad the lifeboat crew would not have been able to moor at their anchorage so they came alongside and were welcomed aboard for refreshments.

Yet still it was not over! When it was time for them to go we waived the lifeboat men off, raised the anchor and moved towards Pin Mill but she started to settle. Our stern end got lower and lower with water begining to rise above the galley ceiling as it ran aft from the fore end. It was back to the pumps all the way back to the hard where we took the ground." John washed and went ashore to contact Peter Boss and Mike Taunton, the Club Secretary to report again and make arrangements for the coach which was waiting for them parked in Ipswich Docks to drive down the coast to Pin Mill. In very hot sunshine that Friday afternoon a tired mud covered group of Scouters finally made their way home leaving a shipwright from PUDGE to make a new steering shaft block, sort out the pumps and tidy up the gear. John later heard that a treenail had come adrift just above the waterline in the bow but his charterers had their picture in the paper on 5th August, 1982.

There were no problems on Marilyn Buekett's Charter from 30th August to 3rd September, the following charter by Peter Ferguson with Peter Boss as skipper, or subsequent club weekends. There was, however a note that on the passage from Ipswich to Greenwich on 13th October CENTAUR was overtaken by an elderly lady walking along the sea wall at Holland on Sea in wind against tide conditions. In view of the southerly gales forecast Peter and Rob called it day and made their way back to Maldon Quay to end the sailing season for 1982.

After the restoration of the bow the next notable 'event' in the CENTAUR story was a severe leak caused by a fastening falling out of her bottom. That resulted in the bottom of the barge being re-spiked on the blocks in Ipswich. Peter was at a management meeting at Fords on 29th June 1983 when a note of a telephone call from the Ipswich Port Authority was passed to him. "Your barge is sinking". Peter sank too, but a couple of telephone calls assured him that one skipper of a barge lying nearby had got a pump on board, another supplied the petrol. A third skipper took CENTAUR to Pin Mill while a fourth came aboard with his crew to help trace the leak and with other Club members overhauled the engine and clean the whole barge of crud which had floated up from the bilge some of which must have been there since 1895, although a crab that appeared in the saloon appeared younger.

Marilyn Buekett, then Club Secretary suggest that the Club should enter the competition for the 'Heritage Award' made by the Conservation Foundation and she was given the go ahead to prepare the paperwork describing our work on the 'Centaur Appeal'. All eyes were glued to the Television screens to view *Pebble Mill at One* on 28th October 1983 for the final programme. We were perhaps disappointed not to have won the overall award but with 600 entries in the Heritage category we did well to win the £2,000 second prize and Barrie Lawrence, Peter and Pat Boss and Marilyn were invited by the B.B.C. to be in the studio for the programme to receive the trophy. It was on display at the Laying Up Dinner Dance at which Marilyn was presented with a bouquet and the guest of honour was Dr.Neil Cossons, the new Director of the National Maritime Museum.

The next winter was spent at Maldon. When the gear was lowered down one of the rigging chocks crumbled so the £2,000 "Heritage Award" was put to good use in replacing both chocks. She came out for another seasons sailing which included a cruise up London River to the Thames Barrier Opening on Tuesday 8th May, 1984. This was a charter arranged by Marilyn Bueckett with Geoff. Harris, one of our members who had become a skipper, in charge for the trip round from Maldon. They started with navigation lights lit at four on Sunday morning sailing round to Grays for the night and going on to the Trafalgar Tavern where they tied up to the rubbish barge just below Greenwich Pier.

Next morning the barge had a visit from the Port Health Authority launch and set off downstream to secure to a lighter at 12.15 to await the celebrations. The bunting was brought out and the barge dressed overall before the instructions came through that CENTAUR should anchor between the bawley NELLIE and H.M.S. BOSSINGTON. There was then a hours cold wait in a north easterly force 5/6 during which the crew took turns to warm themselves at the stove and think themselves lucky they were not among the school children freezing on the open decks of the moored trip boats. Eventually the Queen arrived at the pier, declared the Barrier open and all gates were raised closing the river until 21.30. That was it! The crew took down the bunting and raised the anchor to leave Woolwich for a couple of hours sail downriver to Grays for the night. Some of the crew left the barge then leaving the rest to make a leisurely way back to Maldon with the mate, Terry O'Sullivan organising them in various maintenance jobs like renewing the bottom boards in the barge boat, scraping the missen, painting blocks, etc. etc. etc. Other skippers taking her that year included John Woodman, Bob Wells, Roger Beckett and, of course Peter Boss.

Over the winter of 1984/5 the wooden keelson was chopped out and replaced with a steel girder which members manhandled aboard in sections. This then still left the 'middle bit' that is new sides and decks between the fore and main horses. The only thing which prevented the Club from instructing shipwrights to start work was the small matter of finance!

One of the three sections of keelson being manhandled aboard before being welded together, 1984/5

While Peter Boss and the Club committee pondered on the problem of financing the next stage of her renovation which we thought would cost between £20 and £30,000 on each side, CENTAUR continued to earn her keep with club sailings each summer weekend. Steve Radclife, our Sailing Secretary was responsible for organising our weekend sailing rota and tried to get the right balance between new and experienced members in each weekend crew. At the start of the 1986 season the duty mates were given a new looseleaf Logbook on A4 paper to record the voyages of our barge inscribed with the message "To boldly go where no man has gone before". Whether the Club has lived up to these aims the membership must decide, but there follows a summary of the season.

The first entries in the log were for the usual 'shakedown cruise', a Maldon to Maldon day trip on 19th April under Peter Boss. The second was like unto it, a club weekend with Rob Dudley skipper and Martin Phillips sailing from Maldon to Osea Pier with a party of six members. The log shows them under way at 12.30 and five minutes later they were back alongside to remoor PHOENICIAN which did not appear to be moored to anything at all. They put out springs, stern and breast ropes to secure her with a bow line led back onboard for them to pick up on their return. Then off again to anchor off Osea Pier at two in the morning and get their heads down. In bright sunshine and a light easterly they were under way and started to tack down river at 9.30. They were still at Osea at 10.30...... and at 12.30 when they again anchored for lunch to ride out the rest of the flood tide. Two hours later they set off again in a fresh force 4 and anchored off Brightlingsea. It started to rain at 8 pm and they did not attempt to go ashore. The log notes '9.30 Abstinence!' Next morning they were under way at 8.30 in a misty wet force 2/3 and the only note in the log that the skipper went below for breakfast while the mate got lost but at one o'clock they had apparently found Osea Pier again, started the engine to return alongside at 2.15

Both CENTAUR and PUDGE remained alongside Maldon Hythe on Monday 5th MAY 1986 for an Open Day which turned out wet until well in the afternoon but we had a good number of visitors. The 'Jovial Beggars' charter organised by Bob Bryan tried to get away on 10th MAY when the forecast was south westerly 5 -6; 7 - 6 later. By ten o'clock it was doing just that so they lay at anchor all day at Stone, only a few miles downstream and made the best of it with members of the charter party entertaining themselves and the crew with singing and monologues, folk, clog and Morris dancing. They were joined by PUDGE for an all night party and also gale bound nearby were XYLONITE, WYVENHOE and DAWN. Things did not improve on Sunday so Peter and Martin struck the topmast and got under way back to the quay.

The following weekend John Woodman was skipper with Barrie Lawrence, mate and Keith Gilbert 3rd hand on an attempted passage from Maldon to Hoo but this had to be abandoned as it blew a South West Gale. Later in the week Peter Boss sailed her across with Barrie Lawrence as mate, Keith Gilbert 3rd hand and Gordon Weddell staying over as crew to arrive at Hoo. 'OK?', is all that is revealed about the crossing. On 23rd MAY Peter with Terry made the passage to Greenwich under motor slipping the buoy at 05.00, setting the topsail off Stangate but the log is virtually silent save to note 'moored at Greenwich Pier" at 15.05

Robert Lister had gone around Greenwich putting up posters and Liz Houghton and Wendy Yates hoisted a large banner in our rigging to advertise that our barges were 'Open'. Colin Chambers drew his cartoon posters which were put up around the pier and out round the town as well as on board and our caterers prepared for the visitors. Peggy Woodward and Pam Elkins who looked after the stewards and Peter and Margaret McLaughlin brought and displayed their paintings. Wendy Yates and her team prepared a culinary miracles feeding over a hundred people at the Fitting out party held on board both barges on the Pier and thanks to Frank our publican member we had a generous donation from the brewers Greene King. By popular request Shanty Crew again entertained the company during the evening with their singing. Sunday was the day when half the country were being energetic on behalf of "Sports Aid" and to publicise the event we were jointed on the pier by Simon Le Bon's boat DRUM whilst out on the river was VIRGIN ATLANTIC CHALLENGER II all of which caused no little excitement plus additional spectators .

View aft from aloft

CENTAUR must have got back to Hoo as on 31st MAY 1986 she sailed in the Medway Match with Peter Boss as Skipper, Terry O'Sullivan Mate and Keith Gilbert third hand and eight members including Roger Newlyn and Colin Chambers. They were first over the line at 09.47, passed Garrison at 10.19, and were still in the lead at the outer mark at 11.58 with OAK second followed by JOCK and SIR ALAN HERBERT . However on the windward leg back into the Medway OAK went past and at around 13.00 CENTAUR lost the topmast when, it was believed, the cross trees gave way. It broke just above the top of the mainmast at a point where there was a ring of large knots protruding which would have made it a weak point. The port cross tree was also bent but they could not be sure if it bent before or after the mast broke. So we never knew the real cause, however, no one was hurt and there was no other damage. The mate, third hand and crew did a great job clearing up the debris while Peter sailed the barge back into the Medway, racing stumpy rigged over the finishing line. CENTAUR arrived back at Gillingham around six that evening and moored on buoy. The crew removed the stump next morning to get underway at one and sailed back to Hoo

On 7th & 8th June there was a passage under Rob Dudley with John Conroy as mate and the following weekend CENTAUR followed the Blackwater Match with Mr.Gilbert on board again as third hand, Peter as Skipper and Martin Phillips mate with a full crew including Andrew Phillips, Mike and Pat Collins, Pat Boss and Roger Newlyn. She was still sailing stumpy rigged the following weekend but a 46 foot pole with a suitable diameter of mast quality was finally found at Erith in Kent and transported to Maldon for Cooks to shape up. We had hoped it would be ready in time for the Pin Mill Match on Saturday 28th June, but although the shipwrights did their best it was not until the Thursday before the match that it was ready. Working parties turned up to work through a showery cold afternoon to sort out the rigging but the port after shroud, and backstay were caught under the hounds and it took some while to set the new topmast and sort out the rigging.

The barge was back together again the passage from Maldon to Ipswich on 28th/29th June with Peter skipper, Rob Dudley mate, Keith Gilbert again 3rd hand with Martin Phillips this trip sailing as crew with Pat Boss, Daphne Mair, Colin Chambers and others. Unfortunately the wind was against them from Maldon to just past Walton on the Naze and they had to motor and motor sail. Once in the Orwell they had some good sailing with force 5 - 6 easterly as forecast and anchored off Pin Mill. Although they arrived too late to enter the match, they were there to take part in the festivities during the evening. This was the 25th Pin Mill Match and it was appropriate that MAY was the overall winner and winner of the bowsprit barge as she was the only barge present to have taken part in the 1962 match. On the Sunday the crew had another session sorting out the new topmast and had a quite exciting sail back out again into open sea, all in all David Stern who kept the log noted, 'a very good week end'.

The next club weekend was a midnight start out of Ipswich Dock on Friday 4th July under Bill Polly and Tony Finbow, sailing round into the Stour on Saturday and back to moor alongside the berth in the Dock at midday on Sunday. Rob Dudley took her on a five day charter from 7th to 11th July 1986 with Martin Phillips as mate and Dick Dawson as unofficial third hand. The comments in the log are "Torture, pure torture, where to next year Mr.Dudley?" They sailed out of Harwich Harbour and tacked up past the Deben in a westerly 3/4 encountering XYLONITE who they engaged on the radio. - C. to X, "where are you for"; X to C "Orford and Aldeboro, what about you"? ; C to X "Snap" ; X to C "Repeat Please"; C to X "Snap"; X to C "Where's that? I don't know it" ; C to X "We're going to Aldeboro' too you pillock" ; X to C "Oh... out". Then they beat back to the Ore and entered the river under sail. The log commented "ran aground, sailed off (Mate distracted)" but they then anchored in mid channel. The following day they shortened up in light airs setting tops'l and main and rigged a kedge from stern, which was not needed, and got under way at ten to sail up the river passing MIROSA to reach Orford at Eleven and ran aground! There appears to be some dispute about this but at noon they anchored for lunch by the Martello Tower at Aldeboro' raising the hook at five to get under way under power to anchor an hour later above Orford in a force 4.

Disaster! The log notes the scotch bottle was observed to be empty, the mate in tears (not true he is reformed really!) Then followed a thunderstorm, and a very wet mate needing windscreen wipers nearly missed the pub. The skipper announced an early start and received the usual polite mutiny: next morning breakfast was at eight with the skipper in tears, but the day was spent at Orford and Dick Dawson purchased a new bottle of scotch. The crew took lunch ashore in a fish restaurant but they were under way at three and after discovering the impossibility of sailing out of Orford started the engine and motored out. The log meticulously details these events and the departure of the shore party at nine thirty. They returned with yet more whisky after discovering a delightfully grotty pub that did not close until gone midnight (they didn't get there until eleven) then totally sober (the log specifically confirms this fact!) they made telephone calls while other members of the party tried to overturn the telephone box with the caller inside it. The return trek was very interesting, they got lost and found themselves in a field of bulls. Such is life like on a charter!

Next morning, (Thursday) " Oh Gord, I feel awful", "No, never again!!", but someone was up, and sober, the engine was on at a quarter to ten and with a fair breeze for Harwich the party had a excellent sail into the Harbour and up to Wrabness to anchor off Ewarton. The log notes that the shore party were stranded; no one told them about tides. The mate showed exceptional valour and successful effected a rescue mission to deadening and resounding applause. The galley crew scored a direct hit on returning boat with sink pump! Why do we always land at Ewarton at low tide?? The boat returned with more scotch, only a small one. Next morning the last day of the charter the skipper was feeling remarkably well while the charterparty attended to breakfast and solving the worlds problems like how to get the plastic toy out of the cornflakes packet but by eleven they were under way under sail. Facing a head wind and making no progress they dropped sail, started the motor to arrive back at Ipswich and tie alongside the lock at three. Engine off, I've survived, Thank God its over.

Next day Peter Boss with Doug Nicholls as mate and the erstwhile Keith Gilbert took away a party which included Ronny Lavender and Peter Roberts to Wrabness, followed a week later by Rob Dudley and Keith Gilbert on a weekend attempt to visit Hamford Water which had to be abandoned although they had the pleasure of sailing in company with the gaff rigged SKUA. Then on 24th/25th July, Peter Boss with Terry O'Sullivan as mate sailed from Ipswich at three in the morning to cross to Queenborough where they tied to a buoy at four thirty in the afternoon to rest and recuperate before going on to our berth at Hoo the next day. John Woodman and Rob Dudley had a couple of pleasant, if uneventful, weekends in Kent before Peter Boss took charge again for the Swale Match on 16th August, 1986 with Tony Finbow mate and Keith Gilbert third hand.

The engine on at ten but apart from a note that they gybed off the Spaniard at 12.10 and crossed the finish line at 18.45 the log has little to note. However, in his note for the Bulletin Peter wrote "this was the first real chance to put the new topmast to the test, we made a good start and got the cup for the first over the start line in our class and headed off towards Herne Bay. It was one of those sunny days when the force three south west wind required the stays'l to be set, but the passing cumulus clouds caused the wind the gust, at time to force five. Peter spend most of the match watching the mainmast with his heart in his mouth and dropping the stays'l when the wind pipe up. That stays'l was up and down more times than the proverbial fiddlers elbow. On the return leg while tacking back into the Swale and trying to keep clear of the minefield of lobster pots, CENTAUR suddenly reduced to about half speed and refused to answer the helm. ENA and DECIMA went past us as though we were anchored. We had obviously picked up one of the lobster pots and were dragging it along the bottom after us. We searched for it in vain expecting the float to bob up each time went about. We only found the culprit when it was pointed out to us by another boat after we had crossed the finishing line. The float line had stuck in the end of our chine doubling at the bow, just aft of the anchor and the float was tucked right under the round of the bow." CENTAUR finished the race in third place to OAK and DECIMA which Peter supposed wasn't too bad considering the self imposed handicap. There was an unconfirmed rumour that our skipper had lobster for tea as a consolation.

The Southend match followed the next weekend with Doug Nicholls as mate, and a full crew including Garry and Barbara Butler, Squire and Alan Patchett. With 14 barges competing the match ran out of wind by afternoon and most competitors tacked back and forth all afternoon making little headway over the tide. So the race was shortened and then finished at five with CENTAUR taking 3rd in class C behind JOCK and SIR ALAN HERBERT. On the Sunday CENTAUR made a passage to the Stour in a fresh northerly breeze with a very lumpy sea and sailed into Ipswich Dock on August Bank Holiday Monday just in time before the tail end of "Hurricane Charlie" struck on the Tuesday and Wednesday, which brought some of the worst weather of the Year. Peter sailed into the dock and berthed without using the engine, a feat which caused some comment and a photograph in one of the Ipswich papers.

Hurricane Charles blew itself out though the weather continued very unsettled thoughout all August and September. CENTAUR stayed based in Ipswich and Rob Dudley was successful in tacking up the Hamford Water to anchor there for lunch on 6th September although the following couple of weekends Bill Polly had an even wetter time with continuous rain for some thirty hours over the weekend 13/14 September. He and Keith Gilbert as mate did not get out of the River in the light airs which persisted until Peter brought her on passage back to Maldon for the remaining two weeks of the sailing season. So ended a typical sailing season. Our crews had taken to making their own comments in the log book (one reason for the report on this year) and next season the club committee decided that a Crews Log Book should be kept on board as well as an 'official' log.

So the Club settled into another winter in Maldon. Despite being virtually cut off from the outside world by snow, work in the spring of 1987 was carried on under a tent made from a large boat cover by the light of a couple of fluorescent lights after dark. The restoration of the skipper's cabin roof and quarterdeck was in hand and Brian Gleghorn laminated the new mizzen mast support beam from five planks of Iroko. The traditional carved front of the old beam which has cracked the previous season was kept and glued to the front of the new beam. Tim Jepson was still happily caulking the seams until midnight on one Saturday so the polysulphide

filling of the new cabin roof could cure overnight any we would not loose and time on Sunday. The decking under the main horse was replaced and in the saloon Mike Collins took out the locker seats and replaced them with some church pews in the belief that barges inspired almost religious

devotion.   Two brand new gas cookers were donated by Ronnie Lavender and installed in the saloon.

Ipswich Dock in 1987 had altered considerably in the five years since our last Open Day there.   The Home Warehouse was now a very up market office development, the old Maltkiln, a pub which did good food, (marred however by the usual loud music). Whitmore's sail loft had become a chandlery and no more did trains clank along the quayside under the gantries with their loads of grain.   The river had also changed with the great new Orwell Bridge spanning the upper reaches, a new container base dominating the nearby riverside jutting out into the deep water channel. The people of Ipswich, however, seemed to appreciate another barge in their Dock and the Open Day on 21st June produced a profit of almost £400. Miss. Aline Williamson of Woodbridge gave the club a set of water colours sketched in the early 1960's; later some were sold in aid of the CENTAUR Restoration Fund by Tony Osler well known artist and leading light in the East Coast Sail Trust.   Another member Malcolm Barrell, arranged for an Oak tree to be given to the Club for repairs to CENTAUR.

We still had to raise the money to complete her restoration so CENTAUR crossed the estuary for a charter by a film processing company for their executives.   While she was alongside Strood Pier on Friday 17th July, 1987 a dinner was provided by outside caterers and held onboard.   Rain was falling and tables were set with a bar set up just inside the saloon while Tony Finbow, Brian Dawson, Keith Gilbert and skipper Peter Boss served drinks (and moved the bedding around to keep it dry) while the charterers got up to the sort of things charterers do.   The party slept aboard and next day followed the match, CENTAUR motor sailing to keep up with the fleet for the benefit of the photographers.   At the end of a tiring day the party were put ashore at Strood in the evening and CENTAUR returned to Hoo to swap the old topsail from PUDGE for her own which had been repaired after the Blackwater Match gale. Peter then took CENTAUR to Harty Ferry for the Swale match on a lovely morning with a n.w.. force 3 breeze and came third after GRETA and OAK with ENA and DECIMA close behind

The Open Days on Greenwich Pier in the seventies were very popular and we regularly used to get around 2,000 visitors over the weekend but the number dropped to about 500 on the last occasion.   Visitors seemed to be intent on doing the Museum, CUTTY SARK and returning to their tourist trail with no time for a diversion onto our barges. The Greenwich Open Days on 4/5th June 1988 had to be abandoned as the Pier was removed for repair having been thumped by a passing ship and both club barges moved to London Bridge City Pier alongside Hays Galleria shopping arcade.   Our barges passed under Tower Bridge together with their motors running as the wind was on the nose and quite squally.   We shared the pier with the Thamesline passenger service waterbuses which had started running a day or so earlier. There was a certain amount of flack overheard on the VHF as the service had not proved popular with other river uses including the crew of ENA who were thrown about by wash created by the catamarans but our visitors were in at the beginning of an enterprising but ill-fated new service. The public did not come in the droves we had hoped but our members came for our 40th Anniversary Celebration and Supper on CENTAUR. After the event our barges sailed for the Ipswich Festival of Traditional Sail which was also somewhat disappointing as none of the barges competing in the Passage Match were able to lock into the dock until Monday morning.

Our lighter, SAILORMAN, which had been moved to Maldon from 'the Royals' was given a good clear out at the start of the 1988 winter to find room for over five tons of oak, cut into sizes for frames, deck carlings inwales and lodging knees. Barbara Butler had negotiated with the authorities at Bourne Park Ipswich for the gift of two oak trees which a miller had then cut into the sizes needed. A crew to load them onto a truck which Nigel Martin had borrowed to deliver it all to Maldon and with Barbara and her husband Garry with some help from Paul Curtis, Cathy Lockwood, Janet Stephens and Pat Boss the five ton of timber was lowered into the lighter and stacked for seasoning; a job more suited to a gang of big navvys and there were some aching shoulders next morning.

The engine in CENTAUR was not a proper marine engine but a Ruston designed originally as a stationary engine used mostly in cranes. This one had been marinised and spent some years in a trawler before being installed in the barge. Three and a half knots is all it would do and that's without a head wind. It had become quite temperamental, it leaked oil and we couldn't get spares. The Club had been given a couple of large Bedford truck engines by General Motors in about 1985. They had been run for test purposes only and could not be sold as new so Chubb Chambers who worked with GM sweet talked them into donating them to the Club and had kept them in storage calling one *Bluebell* and the other *Elizabeth*. The engines had still to be marinised and installed; that meant it had to be fitted with a marine gearbox, heat exchangers, pumps, filters, pipework for the cooling water and oil, a new exhaust system, electrical wiring and operating controls and the engine had to be modified to suit. We also had to acquire a new propeller with driveshaft bearings as very little of what we had in the old engine was suitable.

The job was put in the hands of a Maldon marine engineer with our members stripping down and removing the old Ruston and gearbox to reduce costs. The engine room bulkhead was removed and the engine lifted out on a chain and tackle suspended from a wire at the sprit end and we set about cleaning twenty five years of oil, grease and general crud from the engine room. CENTAUR was then put on the blocks to remove the old propeller and shaft, both of which were badly worn and fit only for scrap. The bearings in the old tube had to be removed and sent away to be bored and sheeved to accept the new shaft and propeller, the old engine bed bolts were badly corroded so we got the shipwrights at Walter Cooks yard to knock them all out and replaced. At the same time the engine bed was raised by six inches using the new oak. When CENTAUR came off the blocks a crane tug came alongside the lighter hauled out the new engine, lowered the new one into the galley area and it was manoeuvred into the engine room and onto the new beds. Then the bulkheads and hatches were replaced and Ian Hiner, the engineer set to work. It was not until the 3rd May that we were able to get back on the blocks to fit the new stern tube propeller and skin fittings and a few more days to complete the installation work to carry out trials so she could be operational for the weekend of 14th May, 1989.

"Bluebell" had her first good try-out during an unscheduled midweek passage from Maldon when we had to push CENTAUR into the Blackwater ahead of a late forecast S.W.gale and she performed very well for power speed and manouverability in the very windy conditions. It was now such an advantage to have a good reliable engine which could be left silent while keeping the barge sailing in crowded waters congested with moorings knowing that it can be started at the touch of a button if the barge was in danger of touching a buoy or another boat. Peter Boss badly twisted his 'bionic' knee just before the 1989 Medway match on 27th May and could not sail so had to ask Stan Yeates to sail his old command. Five barges retired due to weather conditions in the strong force 4 - 6 north east wind which caused a nasty swell but Stan got the Seamanship trophy; CENTAUR gained 3rd place to THISTLE and ETHEL MAY. Stan and Chick had yet another claim to fame when they were interviewed on board while at Maldon Quay on Monday 3rd July.

At the end of June 1990 we heard that our berth at Hoo on the jetty made of sunken concrete lighters had been closed as part of plans for redevelopment of the Marina. This left us in a quandary about our sailing season with nowhere to moor on the Medway but Barrie Lawrence and Peter negotiated an agreement to use the old ballast wharf behind Hoo island. This had been unused for many years and two Sunday working parties were set up to clear the berth of old lorry tyres, concrete slabs and with the aid of blocks and tackles and a lot of sweating and heaving we finally got the berth clear enough to moor the barges. At the end of end of May CENTAUR made the passage from Maldon to tie up under the high concrete dockside and hosted the Hoo Ness Yacht Clubs small boat rally in Stangate Creek.

Roger Newlyn seated on the rail

CENTAUR spent the summer chasing from one match to another. After the Medway and Pin Mill it was the Blackwater. The wind was blowing force 5 just before the start and when we set the tops'l the clewline headstick block got caught round the topmast backstay and by the time we had cleared it there was a tear in the tops'l. The five minute gun went so we were not allowed to use the engine. However, we were still a way off with a crowd of small boats anchored between us and the start line in the narrow part of the river with the wind blowing hard. In view of the risk of damage the engine was started and we hauled down our race flag.

The following week was the Swale Match with Rob Dudley as skipper and although the wind died away we eventually rounded the mark to finish third behind ORINOCO and the LADY OF THE LEA. Two weeks later the Southend Match was threatened by thunderstorms which sent the wind round the clock and Stan Yeates, the officer of the day had a difficult job to set the course. CENTAUR was first over the line and ran down to the South Shoebury with the wind which then died leaving her to tack up to the West Leigh Middle Buoy. The match officially ended at five and the only barge in the slow class to get round was ORINOCO. A storm came just before the match ended and the crew got very wet, but after all, wasn't that what we all expected! It was all part and parcel of sailing the Club barges.

But there was still the 'middle bit' of our barge which needed work done on it before she could be said to be fully restored. The world was also changing; rules and regulations were being made to rule and regulate our sailing by far off bureaucrats and we could not disregard either the passage of time or the passage of legislation.

8. The Last Push to the Centenary 1990 - 1995

The barge fleet were thrown into turmoil by something called "The Merchant Shipping (Load Lines)(Exemption)(Amendment) Order 1990" which was laid before Parliament on 7th March to take effect from 28th March 1990.   It was, to most readers unintelligible gibberish, deleting reference to 'sailing ships' in an order of 1968; but had the effect of confining to port most of the barge fleet except MAY, CABBY and the few which already had load line certificates.   The order had been made without any of the usual consultations and was, it seems, directed at a loophole which allowed new cruise liners escape safety regulations by setting sails. It meant, however, that the Maldon charter barges would now be confined to 'smooth waters' in the Blackwater,   while London based craft could venture only half way down Sea Reach.   The Club believed that our barges remained private pleasure yachts not affected by the new order,   but that any more 'commercial' charters to raise money were now ruled out of order.

After representations to the Department of Transport by the R.Y.A. and S.B.A. along with the Association of Bargemen, the Department ruled that the charter barges should be exempted from the new Regulations within 15 miles of the 'partially Smooth water limits' and within 3 miles from the shore with an extension of the area to allow access into the Deben.   The club obtained an Exemption Certificate on condition that our barges carried not more than 12 passengers, were maintained in a seaworthy condition and carried adequate safety equipment including efficient radio equipment.   We were truly grateful to Roger Beckett, Richard Walsh and Peter Dodds who negotiated the exemption but the experience showed the need for the Club to belong to a wider Association.

So we continued sailing during the summers of 1990 and 1991. After the 1991 Open days in St.Katharines Dock and the summer supper with a bar on board CENTAUR, she attended the Medway Match getting second place in the slow staysail class.   She then made the passage to Ipswich and was present in Shotley Marina for the Classic Boat Festival on a members' charter.   A week later she was in the Pin Mill match sailing in a light to moderate westerly breeze on a pleasant sunny day to gain third place. This was followed by the Blackwater and the Swale Matches and then the Southend when Peter Boss decided we were better off staying at Hoo in view of the force 9 gale from the south west which blew on Friday. The wind moderated on Saturday and CENTAUR made her way to Southend to join other barges crawling out of their hidy holes. The force 4 wind died away by midday leaving all the barges down tide without a wind so the prizes were awarded to the barge position relative to the finishing line at the official finishing time and CENTAUR was second to LADY DAPHNE. CENTAUR had another T.V. appearance in the BBC series Lovejoy episode "No Strings" with Joanna Lumley appearing at the wheel and having lunch outside the Jolly Sailor. The club received £250 for the 2 minute slot which actually took over half a day to film and Peter had his day with the stars!

In May the headmistress of Darlinghurst School in Leigh gave a couple of dozen of her pupils a taste of life afloat and the story and photographs appears in the Evening Echo. Pat Boss (for it was she) said "It was a tremendous experience for them to be away from home, they learned to co-operate in a group as you would in any sailing ship. It was quite an exciting adventure for them" But in was not all fun, the trip was planned as a field study in Maldon with the barge as a base to learn in depth about a new area. The pupils visited Northey Island, the site of the Battle of Maldon and did work on the environment of the town. It was the first children's field trip based on CENTAUR and our first venture into a field in which the East Coast Sail Trust have been active for years.

Our first Commodore, Frank Carr died on 9th July, 1991 at the age of 88. Although he was at the helm when the Club was started he was always adamant that the idea had come from Hugh Vaudrey. Although his duties as Director of the Museum soon intervened he later became our President, and a loyal supporter attending our AGM's and dinners accompanied by his wife Ruth, herself no mean sailor. His last appearance at a club event was at the official launch of a new edition of his book Sailing Barges for which the Club Bulletin Editor had contributed new chapters. This was held on board CENTAUR at Maldon in the summer of 1989 with Bob Malster, Terence Dalton's General Editor who had done so much to oversee its production.

By the winter of 1991/2 the Club realised that it was eleven years since the bow and stern sections of the hull had been completed. Over the years we just hadn't been able to get a fund together to pay shipwrights to work for a whole winter and complete the job in one go. The money that we had saved we spend just as fast on essentials, for example the two new engines, two new mainsails, two topsails and one foresail which alone came to £25,000. There had also been a considerable amount spend on PUDGE, and we had reached a stage where we realised that we could not wait for the money to appear. We did not have the time to do all the work ourselves in one winter but some work could be carried out in stages over a period of years. Peter laid plans to replace all the deck carlings, the inwale frame heads and linings and all the deck planks along the side decks. The outer planking outer wale covering boards and deck rails could wait for a year two or three.

Tim Jepson caulking the side deck

This was only practicable with the increasing expertise of the members, particularly Peter himself, who had recently taken early retirement and Tim Jepson who had built his own half size barge some years ago. Tim had recently moved to Maldon and lived near the quay making it possible for him to spend time and advise on the shipwright work. Professional shipwrights can spend 5 or 6 days a week to complete a job well within one season but most of our members have other commitments on their time, like earning a living. However the Club had got to the stage when we had no option, we just had to get on with it ourselves, even if it meant that we might be a little late in completion.

We began by buying a whole Iroko tree trunk over thirty feet long which had been sawn into two inch thick planks. Five of these planks were manhandled by a huge gang of members and stowed on the port deck. These were for the twenty inch deep double thickness inwale. We bought a load of old scaffold poles for a tent frame and built a huge tent over the whole length of the barge from stem to stern. The next move was to put her on the blocks. We were then ready to begin. Before anything was removed we fitted two chains in tension right across the saloon to ensure that she kept her shape, then we totally cleared out everything down below on the starboard side including all the bulkheads cabins galley and fuel tank. We painstakingly cut out all the old inwale and lining using a chainsaw, crowbars and lots of muscle. Many of the frames were rotten for the top two feet where rainwater had been constantly running down through split covering boards into the endgrain of the frames. The rotten frame heads were cut off and new pieces fitted. Full length doubling pieces were then fitted alongside the whole lot being bolted together and through the sides of the barge. The doubling pieces are 6 x 5 inch baulks of oak approximately seven feet long All old wood exposed and new wood fitted was soaked with liberal coats of preservative at each stage of the operation.

The old angle iron deck carlings or half beams, eleven in total had straps going down the barge side and bolted through the frames. There were two coach screws going up through every deck plank and the inboard end was fitted with a 22 inch long bolt which went up through the hatch coaming. Every other one was removed. We bought 2 1/2 inch angle iron in five foot lengths rolled to the camber of the deck and welded straps and gussets on to each one, drilled fixing holes a total of 19 on each and installed them. We then removed the other half and did the same!.

Next came the inwale or beam shelf which is one of the major longitudinal strengthening components going the whole length of the barge. Consequently we had to ensure that all sections are tied into each other. After the planks were cut to `suit the exact shape of the sheer of the barge, the hatchboards were removed and the planks manhandle below, held in position and fixed. Boards of this size are extremely heavy and they all took at least ten men to lift and hold while initial fixing was taking place. All the old bolts had been knocked out so we drilled from the outside, through existing holes in the outer wale and frames through the new inwale and bolted them all up tight with new fifteen inch bolts. All the lining was

Replacing the Deck   1991/2

then replaced in pitch pine. This is an excellent wood which had been retrieved from the old pier structures of the Albert Docks. It must have been in the Thames 150 years but it still smells as strongly as the day it was cut. We bought enough of this timber to replace the outer planking amidships and both the side decks end to end. Although new deck planking was fitted when the bow and stern were rebuilt this was twelve years before and all the joints were around the main and forehorse. If we just renewed the midships section of the deck all the movement of the barge would be forced to these same places and we would be faced with a continuation of deck leaks. By replacing the whole length and gluing all the joints with modern glue we hoped to make continuous planks from end to end, giving much greater strength and reducing leaks. After the decks were laid all the deck fittings, davits, bollards, horse chocks etc. etc. had to be replaced, but this was becoming almost routine as was reinstating the accommodation including the galley repainting and fitting out. Still its O.K.. if we keep reminding ourselves "This is just a hobby"

Nearly all the usual attractions had to be crammed on to PUDGE for the first open day of 1992 at Maldon Quay over the weekend 25/26th April while the CENTAUR working party continued work.   The well organised evolutions of a large working party tending the rigging and raising the mast proved an attraction to many of the punters who enjoyed ringside seats on PUDGE.   One barge only had at first seemed a depressing prospect but the reality was different.   Soon the flags were up and the usual cheerful stewards and stall minders started arriving.   Wares were tastefully and enticingly arranged, club items and white elephants were all lined up for sale below while up on deck, rain and shine the tombola stand was subjected to unrelenting pressure by hoards of children and parents all day.   It was the first opportunity for many members to see our new and updated display stand featuring work on our barge (when it was not being dismantled and hastily taken down into the hold of SAILORMAN to keep it out of the rain).   In the afternoon our accordionist and violinist both wrapped up gave a lively concert to a chilled audience.   On Sunday many came on board to shelter from the rain drawn there by the excellent publicity from Essex Radio who broadcast an interview with Peter Boss.   The weather was rather unkind

In 1992 the Club was back to four regular skippers, Rob Dudley, Doug Nicholls, Terry O'Sullivan and Peter Boss.   Geoff. Harris had taken up the offer to sail XYLONITE for the Cirdan Trust as Adrian Mulville decided at the end of the 1991 season it was time for a change.   Not having Geoff. as part of our regular team gave us the opportunity of asking Kevin Murphy from Maldon and Stan Yeates from his home in Kent to come to our aid yet again.

The lateness in completing the work programme for the 1991/2 winter meant that CENTAUR missed the Medway Match but we were ready in time for the Blackwater Match. It was a very pleasant morning with a light northerly breeze veering north to north east. The start was off Osea Island and the course down river to the Bench Head, up the Colne and round the Inner Bench Head Buoy which was the outer mark.   Then back to the Bench Head and home to the finish line off Osea again.   CENTAUR made a good start and was first over the line in her class.   The Blackwater Match, unlike other matches does not have a cup for the first over the start line in each class, but one cup only for the fastest start of the day in all classes. This makes it really worth-while winning but we had to keep our fingers crossed until the end of the day as we did not know how fast the start had been in the other two classes.   Close astern of us came ETHEL ADA, ENA and CYGNET.   We were very soon passed by ORINOCO but gradually gained on ETHEL ADA and ENA and reached the Bench Head then tacked in to the Inner Bench Head.   The course was well planned as by the time we got back into the Blackwater the flood tide had just started and we ran all the way home. The final result was ORINOCO 1st, CENTAUR 2nd , ETHEL ADA 3rd and yes we did get the cup for the fastest start of the day.

Our next race was the Pin Mill on 20th June 1992. This match presented quite a problem for the officer of the day, Mick Lungley. The inshore forecast was north - east 6/7 and there was no way that most of us would take our barges to sea in that. However, Mick devised a course that would allow us to sail in the strong wind conditions without as well being subject to heavy seas. The course was from Pin Mill, down the river, through Harwich Harbour to round the Outer Ridge buoy which was only about a mile outside the lee of Felixstowe Point, then back through the harbour up the River Stour to the No.2 buoy above Ewerton Ness designated as the outer mark then back to the Outer Ridge and home. This was the first match arranged by Julian Ackland the new Secretary of the Pin Mill Club after the retirement of Jack Haste.

CENTAUR was again first over the line in her class, ahead of ETHEL ADA, ENA, LADY JEAN and CYGNET. We reached down the river and through the harbour to Cliff Foot in a strong wind but with smooth water. The last bit between Cliff Foot and the Outer Ridge was a bit lumpy due a wind over tide situation. The seas were almost abeam so we did not slam into them at all, just corkscrewed. As we came back through the harbour we saw ENA rounding the Outer Ridge but it appeared that all the others in our class had retired. By this time the wind had increased so we dropped our tops'l as we watched the bowsprit class, also stumpy rigged heading out again at what looked a tremendous speed. We rounded the outer mark still ahead of ENA and were then told by radio that the course for the B and C classes had been shortened to the Cliff Foot. So we did not have to go out to the Outer Ridge again. There followed a long but 'interesting' tack back up the river to Pin Mill in the narrow channel, still stumpy rigged. All in all a great days sail: and we took three cups, first over the line, first to the outer mark and first to finish.

Then it was back to normal weekend sailing with three weekends out of Ipswich, then the passage over to Hoo for a couple of club weekends. On 31st July CENTAUR again acted as mother ship in Stangate Creek for a gathering of Hoo Ness Yacht Club boats whose crews were invited on board for the evening. We made the passage round the Isle of Sheppey to Harty Ferry on Friday for the Swale Match which is as much a rally of traditional craft as a match and the East Swale was full of craft, barges, fishing smacks and bawleys, old gaffers of every type including a beautiful East Coast One Design, various dutch sailing and motor vessels and even a couple of old steam driven vessels.

CENTAUR sailed from her berth at Hoo at seven on Friday morning and went over to the Blacktail Spit Towers before turning to run back to round the Columbine at half one. We followed ORINOCO into the East Swale to join VICTOR, MIROSA, WYVENHOE and GLADYS at anchor with a new addition to the fleet, the St.Osyth based bowsprit barge EDME, another product of Canns yard due to celebrate her centenary in 1998.   In the evening Peter Boss left in the bargeboat with mate Tim Jepson, 3rd hand Roger Newlyn (plus his wife, Denise then our crewing secretary) and Roy Goldstraw for the Skippers Briefing bouncing and bailing over a very choppy sea.   The forecast for the morrow was N.E. veering S.E.., 4-5 squally showers and thunderstorms. We got it all plus flat calm.

Early on the the race day REPERTOR, LADY OF THE LEA and GRETA made their way down the creek from Faversham and at last Portlight motored out setting all plain sail as she left the creek to sail straight down to the line and, as we learnt later, take the first over the line in her class.   The previous year CENTAUR and the other restricted staysail class barges had a wonderful view of the whole fleet but this year, with our class last away we saw little of the fleet on the outward leg to the Girdler Tower.   The sands either side of the channel had already claimed a couple of victims and a dutch klipper ketch ended her match near the start line over which we passed at 10.59 some way behind ENA who was first across in our class. LADY OF THE LEA looked a very different barge from last year with a new longer mainmast and several new cloths in her mainsail and Don Grover kept her up with us on the outward leg as we lost sight of land through the mist off Ham Gat.   Now the match became a contest just between the Lady, CENTAUR and out in front, ENA.   By lunchtime we reached the Spaniard and now with Tim at the wheel, in the rain, we closed on ENA.   Across on the Kent shore we could see the blurr of a bowsprit barge, perhaps PORTLIGHT with a couple more bowsprits behind with only four barge lengths between them. The mists came down and all there was left was the three of us heading for the five legged Girdler bird perch.   ENA made the mark at twenty to three, less than a minute before we rounded with the Lady close behind.   The sun broke through the mist from above and at three in the afternoon Peter at last acknowledged that we had passed ENA to windward, but then the wind died.   Staysail out and rolling wangs set to counter the heavy swell, the flood started to run up pushing us back up the Swale faster than the intermittent breaths could take us to an illusive inner Mark.

Deserted by all, no sign of a committee boat, no response on the radio we drifted in a half mile circle of sea blanketed mist.   ENA had enough of this, stowed up and motored back towards Harty to drop her charter party.   We consulted the rules and noted that craft finishing after 4 p.m. might have to time themselves over the line and decided to stick it out.   There was no news from the organisers about shortening the course but modern science was at hand. If Essex man Peter could not smell his way over Kentish mud to Herne Bay like our old skippers from working days, his magic number box enabled Tim to translate digits on a screen to fixes on the chart. An occasional southerly carried us inshore and just after four we sighted the orange inflatable K.S.A buoy which was the inshore mark. Then the course was set for Whitstable Street with the echo sounder buzzing as we went over

Studhill. Lobster pots to port to starboard, to port and all round, but we missed them all and now with a freshening wind back to North East we ran back into the Swale. Past the Columbine with the Whitstable Street buoy well to port (as required by the race) rules our staysail running dart the crew left only Peter on deck as we continued our run. The crew had, of course gone below for a curry dinner but by ten to seven we had returned on deck with our pots and pans, trays and bells and the hooter to respond in kind to the gun from a patient committee boat. We anchored; Lady took her gun some half hour later to join the fleet, now including the barge yacht TINY MITE and the unrigged Peter Nicholls built barge yacht FRANCES, at anchor. Then ashore in the boat to Hollow Shore where Richard HughPerks officiated at the prize giving. Class by class, winner, second and third and each contestant as recorded by the organisers until Hugh announced "Restricted Staysail first....ENA" "Oh No She's not" we all cried. Her mate came forward to confirm (skipper Tom Polly being engaged with Terry Everett outside the Shed). So Peter did pick up his Pennant and Pot, the one presented by our Commodore, Barrie Lawrence. At last the barge boat back through the phosphorescence to the barge with the brightest riding light.

The passage of London River started on Thursday morning and ended in locking into St.Katharines around six on Friday evening to join PUDGE after her charter for the Open Days. In St.Kitts were WYVENHOE (with " Welcome aboard Wyvenhoe" notices stratically placed around the dock), FELIX, GLADYS, ARDWINA, BERIC and JOCK. LADY DAPHNE was out as was her sister LADY JEAN (named "Sir Alan Herbert" while owned by the East Coast Sail Trust). The Southend match was due to be sailed on Saturday 29th August 1992 but due to adverse weather there were only nine starters out of an expected entry of seventeen. There were reports of force 10 winds from the second deepest August depression this century at 971mb and gusts of 60 mph. In these circumstances skipper Peter Boss took the decision on leaving the Open Day to sail on past the pier to make the Blackwater and Maldon before the worst of the storm struck.

The winter of 1992/3 was a particularly difficult one for Peter, Tim and the club members of the working parties and the shipwrights who agreed to renew all the outside planking from the outer whale down to the chine. Walter Cooks shipyard was now under the new management of Nigel Cardy with his two shipwrights Baden Dedman and Ricky Cardy running the barge repair side of the business. They agreed to replace the portside frames, inwale, outerwale and outer planking amidships while club members replaced the deck carlings, coveringboard, rail, linings and the deck from end to end as well as doing most of the ripping out of the old linings and preparatory work. All went well until we found that the port chine keelson was rotten and there was nothing for it but to rip it out as well. This timber fifty foot long, sixteen inches deep and seven inches thick was fitted inside along the straight section of the bottom. We bought two thirty foot lengths of pitch pine which were scarfed together for the new chine keelson and had to be lowered into the barge with a crane on the stern of the shipwrights tug along with the twenty seven foot length of Iroko for the inwale. Forty foot of the outer wale was also replaced in iroko before we began to fit all our new steel and wooden deck carlings permanently in place.

There was one consolation, having to renew the chine keelson with the rest of the port side meant we would not have to strip out all the accommodation again when the time came to do the starboard chine keelson. When she was built the oak frames were fitted down into the floors

The shipwrights lift the new chine keelson into position

with a mortise and tenon joint and fastened through horizontally with a trunnel. The shipwrights were able to clean out the old holes, make new trunnels and refit the new frames 19" apart just as they had been originally fitted in 1895.

The new port side outer wale and planking

The removal of the forehorse highlighted rot in each end necessitating another rebuild job and the working party also replaced the rudder pin, rebushed the steering arms and the yoke hinge mechanism taking all the slop from the wheel, so that we wouldn't be able to blame the steering gear for not being able to steer a straight course next season. They also built a new port leeboard which caused much interest to visitors walking along Maldon Quay trying to guess what it was.

The Essex Heritage Trust was founded in 1900 to save what remains of the counties heritage and make it accessible to the people of Essex. The Trust had helped some 40 projects including Bocking Windmill and as CENTAUR was built in Essex, traded there in her working days and still sails the same waters, our Publicity Officer, Val Valentine-Smith, though it worth while the Club applying for a grant. A submission was made and examined by the trustees along with other entries including the restoration of a rare early wall painting in a church at Little Baddow, a dilapidated Grade II barn and some Elizabethan almshouses near Audley End. Following their meeting in June 1993 the Trustees awarded the main grant of £2,500 towards the work on our barge. A presentation was arranged on board on Friday 1st October '93 when a cheque was handed over to Peter Boss on behalf of the club and working parties by Mrs. Beverly Barton, the then Chairman of Essex County Council. The weather was generally appalling but a break in the rain arrived at the right moment to allow the presentation to go ahead as planned.

As both barges were sailing on a club weekend there was a good gathering with some of the regular members of the working parties. Light and liquid refreshments were provided for everyone, the local press appeared to be well satisfied and we signed up a new member, Mrs. Barton.

There was no pontoon available at Maldon in the *Presentation of the Award, Peter Boss with Val Smith* winter of 1993/4 so we had to postpone plans to replace the starboard side chine and chine keelson so it was decided to complete the deck restoration which meant jacking up the main hatch coamings and replacing the kingplanks, the double width planks running from bow to stern under the hatch coamings. On inspection of the mainmast after removal from the barge we found the lower six inches was rotten. In addition, on removal of the crosstrees beds and the trestle trees, we found more rot extending a couple of inches into the mast at the hounds. There was nothing we could do to save it. The mast was very old and has been in at least two other barges before CENTAUR. We explored all the avenues for replacement; have a new one made of steel, buy a baulk of pitch pine and have the shipwrights fashion it - or do it ourselves; or buy a secondhand mast. The previous year GLADYS had a new larger steel mast made to replace her wooden one which was still in good condition and we made the decision therefore to buy that. Although it was about three foot too long so we had to modify it but it was much cheaper that the other two alternatives. It was laying in a yard up river at Fullbridge, so we had to take the barge up to collect it. With the extra work we decided to cut short the deck replacement and just do the forward end back to the main hatch, leaving the rest for the next year.

The foredeck with the hatch and windlass removed

We removed the forehatch coamings and headledges repairing their broken half dovetail joints. The mast - case and the windlass were removed. and all the planks were taken up from the sailing deck and the forward central deck right into the bows. We then discovered what appeared to be a small area of rot in the main wooden beam through the fo'c'sle. This beam forms the main structure at the round of the bow. On investigation the rot went a considerable way through the heart of the beam and extended right under the new planking laid the previous year. It was a difficult and time consuming job to cut it all out and to fashion a new length to scarf into it without disturbing and damaging areas previously restored. But it was completed successfully. The two twelve inch wide King Planks were fixed in position, the planks sandwiched between the windlass bitt knees and the fore and aft beams were in and all the new deck forward of the forehatch had been laid. This completed the most awkward fiddling part of the job. One job that went according to plan was the building of the new starboard leeboard (even if it did mean people working for many hours in the pouring rain).

We then fitted the new planks right across the sailing deck and caulked and paid all the seams. The next job was to rebuilt the forehatch by joining the coamings to the headledges, fitting the hatchboards, hatchcloth and replacing the four cabin skylights. We then got a midweek evening work party together and dismantled the tent. On the next evening we took CENTAUR up river on the tide to Fullbridge, had the new mast craned in the following morning and brought her straight back to Maldon Quay. We encountered a few problems when setting up our existing, newly served standing rigging to the new mast but they were overcome with some minor modifications. She was fully rigged and floated off ready for the shakedown sail and in first class order for the St.Katharines Open Days and the start of the1994 sailing season.

This passed with all the usual events, matches, open days and happenings including the Maldon Traditional Sail Festival where our restoration display panels were on show in Cooks shed. The sailing season ended with the customary 'get together' at the Laying Up Supper held at the Maldon Little Ship Club when our Commodore Roger Newlyn presiding for the first time drew attention to the fine display of match trophies won by CENTAUR in her 99th year.

Over the previous years we had restored the bow and stern sections and the port side, all except the chine. This is the strong hardwood 18" x 4" plank some 60 feet long which is part of the structure holding the side to the bottom on the outside of the barge. This could not be taken out and replaced at the time without putting the barge in dry dock. We now had to renew the same plank on the starboard side of the barge plus all the outside planking on the starboard side amidships, the chine keelson (which holds the bottom to the side on the inside of the barge) and the outer wale (similar to the chine and does the same kind of job, but at the deck level instead of the bottom). Although there is no convenient dry dock at Maldon, the yard owns a pontoon around 100 ft by 40 ft in size. It is towed down river and moored on the flats to sit on the bottom. When the tide goes out the sea cocks are opened up and on the next tide it fills and remains sunk, allowing a barge to be floated on top. Then the water drains out of the pontoon on the next ebb tide, the seacocks are closed and she is floated again and towed back to the yard with the barge on top.

So that is what happened. We took CENTAUR down river to Heybridge on Saturday the 5th November 1994; and towed her high and dry on the pontoon back to Maldon on the Sunday.

Under tow back to Maldon on the pontoon

The shipwrights immediately started work on removing the port chine doubling which we discovered was not original. Baden Deadman, the shipwright who had been working at Cooks Yard for the past fifty years, remembered fitting it sometime around 1946 to cover damage to the lower part of the chine. No one is quite sure when the original damage occurred, however, it is most likely it is that which was sustained at Dover, just before the Dunkirk evacuation. This put us in a bit of a quandry. The doubling extended fore and aft well beyond the ends of the main chine plank and if we just replaced the chine plank we would have to replace the doubling over the top of it. We had a rake around in the forecastle and engine room bilge and decided that the chine extensions had seen better days. So we bit the

bullet, pulled out the whole of the doubling and chine extension planks from the bows to the stern on the grounds that the saving in not having to replace the doubling would go a long way to paying for the extra work and we would be putting her back to her original 1895 condition as built.

On the pontoon, the port chine and extension replaced

Club members again built a tent over the whole barge and with both sides taken out we had a good opportunity to examine her floor frames. She was very strongly built with her floors varying from eight inches to twelve inches wide and all in remarkably good condition. The 50 foot length of the old starboard chine keelson was cut out and removed along with the 3 iron knees. Over the last few years we had replaced the whole of the deck planks with new timber except for the two 12" king planks which fit under the main hatch coamings. Now we had the job of replacing the king planks from the sailing deck to the stern, the forward headledge (the forward, upright end of the hatch) and after headledge and fitting the whole of the hatch back in position. So the hatch covers, skylights, and mainhatch coaming and headledges were all removed! While she was out of the water the underside of the barge was scraped and then pressure washed after being inspected for worn out spikes which were renewed. All exposed timber was treated in preparation for the next one

Peter Boss, clipboard of work in hand

hundred years while all the old wood was carted to the lighter and stowed away for burning.   Of course normal maintenance work was carried out as usual; the worn out anchor chaffing plate was removed for renewal; the toilet sea water holding tank stripped out and cleaned and the usual totally unexpected difficulties came to light and, one way or another were solved.

With CENTAUR already on the pontoon and with her chine planks ready to be replaced, it fell to Peter Boss to report to the January Committee meeting that there was now a world shortage of Iroko, the cost was rising dramatically and during the last month it had proved impossible to find a suitable log.   It was accordingly decided to buy a log of Opepe at the cost of some £3,200, some of which could be sold to the owners of GEORGE SMEED to complete her restoration, two planks cut for PUDGE, and the remainder put in store for future restoration.   The timber was delivered to Fullbridge Wharf and craned into the shipwrights lighter and transported to the yard.   Due to this delay timber work did not start until the afternoon of 20th January, meanwhile working parties laboured on the deck.

For one weekend working parties were suspended to allow the workers, including Peter and Pat to attend the Annual General Meeting at the National Maritime Museum and the Centaur Centenary Dinner on R.S. Hispaniola at her permanent berth just upstream of Hungerford Bridge.   The evening was organised by Val and Rita Valentine Smith, Peter Hearn had brought some of his Centaur Centenary Plaques which were sold in aid of her restoration and a band of musicians played melodies of yesteryear as a background to conversation.   The After Dinner speech was made by the President, Mr. Barrie Lawrence; he thanked the various people who had helped with the restoration work over the years and with the work still to be done during the coming winter.   He drew members attention to the limited edition key rings which had been left for them at their place settings, these had been made by Steve and Wendy Yates from the original 100 year old timber removed from her a few days ago, as good as the day it way laid down by those master craftsmen John and Herbert Cann of Harwich.   Barrie proposed a toast to CENTAUR and the company drunk to the next 100 years.

Barrie had been placed next to a lady member who was a stranger to him but who he discovered over drinks was Mrs.Ruth Clarke, the youngest daughter of Herbert Cann, the builder of our barge.   She and her Bank Manager husband had been heavily involved with the R.N.L.I., returning to Colchester in 1964 to be near her widowed mother, (like Barrie, they were avid followers of Rugby football) and she is still very involved with the work of the Institution, being awarded a bar to their Gold Badge in 1991.   At the end of his speech Barrie brought her forward to cut the Birthday Cake iced in blue with the likeness of our barge in red ochre and black coloured icing.

Mrs. Clarke told us that after he had retired, in 1934 her father built a new house in Barrack Lane commanding views of the harbour which was named 'Alstonville' after the town in Australia to which her brother Harold had then emigrated. Much timber from the shipyard had gone into the new house particularly in building the fine oak staircase and one room, which was panelled and fitted out in the manner of a barge cabin. She recalled that at the back of the house there was a brick with the letters "R.C.C", (Ruth Conway Cann) cut into it and buried beneath, a new 1934 shilling and how on special occasions her father flew the national flag and his consular pennant from a mast in the garden. Mr.Cann had been honoured with the order Chevalier of the Orange Nassau by Queen Wilhelmina in 1938, and the Order of the White Rose of Finland in 1939. In his capacity of Vice Consul he had the task of helping the nationals of those countries, and Mrs. Clarke recalls that after the sinking of the SIMON BOLIVAR the survivors were brought to Parkestone. Her father arranged food, clothing, and hospital treatment for them and they were all caught in an air raid shelter during a raid on the port.

When MAY celebrated her centenary in 1991 Tate & Lyle invited Mrs. Clarke to a reception on board at Tower Pier and a trip under power down to the Barrier. Later when MAY came to Harwich, Mrs.Clarke was invited to take a mystery tour from Halfpenny Pier. Skipper David Copsey cast off and soon set sail stopping the engine to quietly sail towards Mistley, where Capt Charles Stone had lived a century before and away from the bustle of the ferry terminal town. Mrs.Clarke was given the wheel of the earliest surviving barge built by her father: she was now firmly 'hooked on barges' like so many who join the Club. Searching though her family papers she came across some unused company notepaper and a plan of her fathers yard which she copied for the Club archives and the last page of this book. She also made a pencil rubbing of the brass plate from the door to the sawmill in the shipyard where CENTAUR had been built in such terrible conditions one hundred years before.

A few days after the reception, on 15th February 1995, in bright sunny springlike weather with flowers already breaking though, CENTAUR celebrated the centenary of her launch, out of the water on the pontoon. Pat and Peter Boss could not let this auspicious occasion could not slip by without some celebration so CENTAUR was decorated with flags, banners, birthday balloons and cards. The birthday cards were received from Pam and Ted Siggory and from Shirley and David Orchard and lunchtime saw Peter and Pat Boss, Tim Jepson and Shirley with the shipwrights, Ricky Cardy, Baden Dedman and Ian Danskin cracking open a bottle of champagne on board to drink a toast to her next 100 years. As they laughed and sipped champagne they considered how honoured they were to be able to play a part in such an important aspect of our maritime heritage. Their thoughts were with all the people who had 'made it happen'.

POSTSCRIPT
1995 -

In 1955 the Certificate of Registration as a British Ship was surrendered when CENTAUR became a lighter and that might have been the end of the story had it not been for the foresight of Richard Duke, the efforts of his team and latterly of members of the Thames Barge Sailing Cub so that for the past thirty years she has been under sail as an unregistered private yacht. In 1992 she was re-measured and found to be 88 ft from the outside of the stem to the rear of the rudder post, not including the rudder and 19'6" over the rigging chocks or 21 feet including leeboards. She now has a British made Bedford 500 6 cylinder engine, diameter 115.9mm (bore) length of stroke 129.5mm, 500 cubic inches, 8.2 litre 78kw governed down to 1,500 rpm giving her a speed of 9 knots.

In 1993 CENTAUR won the Inter Match trophy. This was inaugurated in 1965 under the direction of the late Fred Cooper and is presented each year after the Southend Match, It is given to the barge with the best overall performance in the four senior matches, the Medway, Blackwater, Pin Mill and Southend and is operated on a points system with points given for the fastest start, first round the outer mark, first, second and third, at the finish in each class. This means that the barges in the slower classes have just as much chance of winning as the very fast barges. It was a particularly proud moment for the Club when CENTAUR won the trophy after 19 years of ownership, not just for the skippers and crews for no matter how well a barge is sailed it can only be to the limits that the condition of the hull and gear will allow.

What is important is that over the years since the Club bought her, all the work that has been done to improve her hull and gear has not only improved her strength and durability, but also her speed, sailing performance and safety. The Club must thank, not only the professional shipwrights and engineers but above all those volunteers who over the years have given up so much of their time, made so many sacrifices and worked so hard to achieve this goal.

One of the objects for which the Club was formed was to educate members in seamanship and the handling of the larger types of sailing vessels in tidal waters. It must be hoped that in the stream of Local, National and European legislation and regulation which is now flowing, future generations of Club members will still be free to continue the traditions of our seafaring nation.

Appendix 1
Barges Built by Cann of Harwich

Parsons and Cann

Year	Name	Number	Tons	Owner	Note/Type
1871	Phillis	67247	65	Philip Gane	
1872	Mecanic	67252	117	John Wells,	Schooner
1873	Cyllene	67255	98	R.S.Barnes	Ketch
1873	Amy	67258	98	G.Cann	Ketch
1874	Enterprise	67260	71	Walter Jno. Watts	Note a
1877	Francis	78971	77	Martin Sanders	Ketch
1877	Florence	74439	69	Luke Richmond	Note b
1879	Yulan	78980	73	Geo Cann	

George Cann

Year	Name	Number	Tons	Owner	Note/Type
1879	Volunteer	81163	63	Richard Horlock	
1880	Eureka	81168	73	Richard Horlock	
1880	Muriel	81174	23	Samuel Groom	Note c
1882	Una	84030	74	John Whitmore	Mulie
1884	Glenrosa	84037	70	John Whitmore	Ketch
1884	Hector	84039	22	Hector Stone	
1885	Excelsior	84040	45	R. Horlock	
1886	Haste Away	86628	49	James O.Fison	
1887	Mazeppa	91330	79	John Wells - Horlock	Ketch
1889	Irex,	91334	69	Robert Lewis	
1889	Freston Tower	95307	50	Wm.Morley Wood	

John Cann later J & H Cann

Year	Name	Number	Tons	Owner	Note/Type
1889	Dorothy	91335	67	R.R.Horlock	
1890	Carisbrook Castle	98104	76	Watkin & Howard	Ketch
1891	Mistley	91336	64	Samuel Horlock	
1891	May	97680	61	John Hooker	*
1892	Swan	91339	49	Robert Brooks	
1893	Felix	97686	68	R.J.Smith	*
1894	Ethel	99453	72	Joseph Holmes	
1895	Centaur	99460	62	Charles Stone	*
1895	Kitty	105418	65	Horatio F. Horlock	*
1896	Beric	105421	63	Joseph Holmes	*
1898	Edme	105425	50	F.W.Horlock	*
1899	Marjorie	109877	56	R.A.Horlock	
1900	Kimberley	109209	65	James O.Fison	*
1901	Gladys	109882	68	Cranfields	*
1903	Resolute	116172	60	F.W.Horlock	
1904	Memory	113758	65	James O Fison	*
1906	Edith May	116180	64	W.T.Bagrey	*
1914	Leofleda	132906	48	E.Marriage & Co. Ltd.	

Note a Enterprise is registered as Parsons and Cann, but considered to be built by Vaux
Note b According to Harwich & Dovercourt Free Press 14 July 1877 both Active and Florence built by
Vaux and were both launched from the Naval Yard
Note c Muriel considered to be built by Norman 1880   Note d   All craft marked * survive to 1995

Appendix 2
OWNERS

compiled from information available on the closed register of Transactions
held at the Public Record Office, Kew under ref. BT110.1339

Owners name, Address & description	No of Shares & date		Disposal, No. Purchaser & Date
Charles Stone, (first Managing Owner)	40	-	16 to Robert Lewis 18.2.'95 8 to Ebe. Whitmore, 25.2.'95 16 to Geo Langley 19.10.'03
Eliza Stone	16	-	to Edward Hitchcock 3.5.'01
Mary Ann Barnett	4	-	to W.E.Rogers 21.10.1903
Caroline Underwood	4	-	to Katie Hitchcock 1.2.1908
Robert Lewis	16	18.2.95	to Geo. Langley 19.10.1903
Ebenezer Whitmore shipbroker of Dunkirk	8	1.3.95	to Harry Barnett
Edward Hitchcock merchant of Lavenham, Suffolk	16	8.5.01	to Arnold Hitchcock 14.1.02
Arnold Hitchcock farmer of Totham Hill Farm, Gt.Totham,	16	14.1.02	to W. E. Rogers 28.8.1911
George Langley (Man.Owner 19.10.03) wine merchant of 82/3 Barrack St. Colchester	32	19.10.03	4 to W.E.Rogers 8.8.1904 28 to W.E.Rogers 3.3.1906
William Edward Rogers bargemaster of 34,Port Lane, later Barrack St. Colchester	4 4 28 28	21.10.1903 8.8.1904 3.3.1906 28.8.11	36 to E.A.Hibbs 19.9.1911 28 to E.A.Hibbs 29.8.1911
Harry Barnett Shipbroker of 52 Leadenhall St. London	8	7.2.07	to W. E. Rogers 28.8.1911
Katie Hitchcock wife of Arnold Hitchcock of Great Totham, Farmer	4	8.2.08	to W. E. Rogers 28.8.1911
Edward Alfred Hibbs Sailmaker of Brightlingsea (Managing Owner 26.9.11)	28 36	29.8.11 19.9.11	42 to various 26.9.1911 22 to Owen Parry Ltd. 1.1.'16
Wm Stokes Jarvis	21	26.9.1911	to Owen Parry Ltd. 1.1.1916
Thos Sydney Parry	4	26.9.1911	to Owen Parry Ltd.1.1.1916
Thos Wm. Watts	4	26.9.1911	to Owen Parry I td. 1.1.1916.
Jas Arthur Pawsey	4	26.9.1911	to Owen Parry Ltd. 1.1.1916
Owen Parry	5	26.9.1911	to Owen Parry Ltd. 1.1.1916
Hugh Wake Pawsey	4	26.9.1911	to Owen Parry Ltd. 1.1.1916

Owen Parry Ltd.   64   1.1.1916   to John Sawyer 21.11.1917
(Managing Owner, Wm.Stokes Jarvis 18.1.1916)

John Sawyer   64   21.11.1917   16 to Wm.Patterson12.12.17
  10,Colne Road   16 to J.Butlin   12.12.1917
  Brightlingsea   16 to E.H.Pudney 12.12.1917
  Barge Owner   16   9.5.23   8 to C.E.Gilders, 14.6.1923
  (Managing Owner 21.11.1917)   8 to J.Francis, 14.6.1923
                                   died 7.10.1933 probate to Exors

William Pattison   16   12.12.1917   to John Sawyer 9.5.23
ship chandler of 31 Sydney St. Brightlingsea

Arthur James Butlin   16   12.12.1917   dies 21.7.1919 Probate to Widow
Master Mariner of169 Frinsbury Rd. Rochester

Edward Harvey Pudney   16   12.12.1917   to Ephraim Cripps 23.8.1923
Master Mariner of 4 New St. Brightlingsea

Alice Butlin (Executor A.J. Butlin) 16   8 to Joshua Francis 5.2.1923
                                8 to Cecil Ed. Gilders 5.2.'23
William Pattison   16   to John Sawyer 9.5.1923

Joshua Francis   8   5.2.1923
  34,New Town Rd.   8   14.6.1923   16 to Francis & Gilders Ltd. 2.10.33
  Colchester, Barge Owner

Cecil Edward Gilders   8   5.2.1923
  St.Osyth Lodge,   8   14.6.1923   16 to Francis & Gilders Ltd. 2.10.33
  Leigh-on-Sea, Shipbroker

Ephraim Cripps   16   23.8.23   to Francis & Gilders Ltd 2.10.33
  2,Osborne Cottages,Ripple Rd. Dagenham, Master Mariner

Fred Sawyer,   Robert George   16   22.1.34   To Francis & Gilders Ltd.21.12.1933
Nicholson & John Fowler, Executors of John Sawyer decd.

Francis & Gilders Ltd.   48   2.10.33
  4/5 Hythe Quay,   16   21.12.33   Certificate surrendered
  Colchester, Essex   12.10.1955
  (Managing Owner 13.12.1951, Paul Camden, 28 Hythe Quay)

****

Extracts from the Port of London Barge Register records held by the Museum in Docklands, c/o The Museum of London, London Wall,

CENTAUR   Register 4 - p.686
PLA Barge No. 13651   Length 85.6; width 19.5; depth 6.3
                     Measured tonnage 59 3/4; Burden tonnage 99 1/2
                     Date of registration 27 Sept. 1928

PLA Owner No. 1702   John Sawyer, Joshua Francis, Cecil Edward Gilders
                     & Ephraim Cripps; 10 Colne Road , Brightlingsea Essex:
                     reg. cancelled sold 27th Feb.1934 sold to Owner No. 687

PLA Owner No. 687   Francis & Gilders 4 & 5 Hythe Quay Colchester, 9 Mar. 1934
                     re. Cancelled, not in use in port 6 Aug. 1936

123

Appendix 3
MASTERS of s.b.CENTAUR in trade

Charles Stone	23.2.1895
Dennis Shine	? 1902
William Edward Rogers	? 1903, 1905 - 1911
Charles Caney	18.9.1911
Ephraim Cripps	? 1914

Details from the original Certificate of Registration of s.b. CENTAUR issued to John Sawyer following a resurvey on 25th March, 1912 showing him as owner. The Certificate was held by the Master of the vessel from time to time and was surrendered to the Registrar of Shipping for cancellation on 12th October 1955 on the vessel being converted to a lighter. It is held at the P.R.O.on file BT110.1339 .

Arthur Butlin (at Whistable)	14.	12.	1917
Harry Pudney (at St.Valery en Caux)	27.	7.	1919
Henry Hollands	4.	9.	1919
Ephraim Cripps	14.	12.	1922
Arthur Keeble	2.	7.	1935
Robert Gosling	3.	8.	1945
Arthur Keeble	24.	9.	1945
Michael Banyon	29.	10.	1946
Francis Robert Hewson	18.	2.	1948
Stanley Yeates	12.	6.	1950
Thomas Henry Victor Humphreys	11.	2.	1952
Frederick McVicker Wilson (Nelson)	8.	5.	1952
Brynley Seton Wrightman (sic)	10.	1.	1955

****

Skippers with Richard Duke

Mick Lungley,   Derek Ling,   Tommy Baker,   Theo. Horlock

Skippers while owned by Thames Barge Sailing Club

Vic Wadhams,	Higham	Harold Smy	Ipswich
Horace Briggs	London	Len Polly	Ipswich
Harold House	Sittingbourne	Bill Polly	Ipswich
Leslie Williams	Sittingbourne	Pat Fisher	Ipswich
Stan Yeates	Higham	Mick Lungley	Ipswich
Donald Grover	Faversham	Dave Wright	Ipswich
Charlie Frake	Faversham	Gerard Swift	Maldon
John Woodman	Gillingham	Bob Wells	Maldon
Gordon Williamson	Ipswich	Roger Beckett	Maldon
Terry Everitt	Pin Mill		
Ruben Webb	Pin Mill	Jim Lawrence	Brightlingsea

Club Member Skippers:-

Peter Boss	Rob Dudley	Geof Harris
Tim Jepson	Kevin Murphy	Doug Nicholls
Terry O'Sullivan	David Ward	

Appendix 4
Trading Routes of s.b. CENTAUR

Details given in the surviving "Half yearly Agreements and Account of Voyages and Crew of a ship engaged in the Home Trade only" held at the Public Record Office, Kew under class reference BT99 The records have been dispersed with 1895 and every 10th year lodged with the National Maritime Museum, a 10% sample (which includes the years detailed below) retained by the P.R.O. and the rest either distributed to local record offices or sent to Memorial University of Newfoundland.

1898 P.R.O. ref. BT99/2050

2 Jan	Dover	Lt	Calais	2 Jan
3 Jan	Calais		London	4 Jan
5 Jan	London		Mistley	15 Jan.
24 Jan	Mistley	Lt	Calais	26 Jan
27 Jan	Calais		London	29 Jan
12 Feb	London		Dover	13 Feb
19 Feb	Dover	Lt	Calais	20 Feb
22 Feb	Calais		London	23 Feb
4 Mar	London		Dover	5 Mar
12 Mar	Dover	Lt	Calais	12 Mar
15 Mar	Calais		London	17 Mar
2 Apl	London		St.Helens	5 Apl
7 Apl	St.Helens	Lt	Calais	16 Apl
19 Apl	Calais		London	20 Apl
3 May	London		Southmptn	7 May
13 May	Southtn	Lt	Teignmth	16 May
23 May	Teignmouth		Mistley	30 May
20 June	Mistley	Lt	Dunkirk	22 June
29 June	Dunkirk		London	1 July

7 July	Holehaven		Antwerp	8 July
12 July	Antwerp	Lt	Newport	16 July
25 July	Newport		London	27 July
12 Aug	London		Colchester	14 Aug
20 Aug	Colchstr	Lt	Harwich	20 Aug
see below				
1 Sept	Harwich	Lt	Dunkirk	2 Sept
6 Sept	Dunkirk		London	9 Sept
15 Sept	London		Antwerp	18 Sept
20 Sept	Antwerp		Southmptn	20 Sept
17 Oct	Southmptn		Alderney	25 Oct
1 Nov	Alderney		London	4 Nov
17 Nov	London		Ipswich	21 Nov
30 Nov	Ipswich	Lt	Yarmouth	30 Nov
13 Dec	Yarmouth	Lt	Lowestoft	13 Dec
16 Dec	Lowestoft		Bruges	18 Dec
24 Dec	Bruges		Harwich	7 Jan

1899 P.R.O. ref. BT99/2104

23 Jan.	Ipswich l	Lt.	Queenbrgh	24 Jan
1 Feb.	Queenbrgh		Dunkirk	2 Feb
11 Feb.	Dunkirk		Yarmouth	15 Feb
22 Feb.	Yarmouth	Lt	London	23 Feb
4 Mar.	London		Flushing	7 Mar
9 Mar.	Flushing	Lt	Antwerp	10 Mar
17 Mar.	Antwerp		Shoreham	19 Mar
31 Mar.	Shoreham		Rochester	2 Apl
11 Apl.	Rochester	Lt	London	11 Apl
17 Apl.	London		Shoreham	21 Apl.
2 May	Shoreham		Rochester	3 May
5 May	Rochester	Lt	London	7 May
16 May	Flushing	Lt	Ternenzen	27 May
2 June	Ternenzen		Dover	4 Jun
10 June	Dover	Lt	Ostend	11 Jun
15 June	Ostend		Lowestoft	17 Jun
23 June	Lowestoft	Lt	Harwich	26 Jun

1 July	Harwich		London	2 July
6 July	London	Lt	Rochester	7 July
10 July	Rochester		Dunkirk	12 July
22 July	Dunkirk		Ipswich	24 July
29 July	Ipswich	Lt	Mistley	29 July
2 Aug.	Mistley		Hull	26 Aug
31 Aug.	Hull	Lt	Goole	31 Aug
5 Sept.	Goole		Southampton	11 Sept
20 Sept.	Southampton		London	29 Sept
5 Oct.	London		Hull	11 Oct
30 Oct.	Hull		Faversham	8 Nov
14 Nov.	Faversham	Lt	Harwich	14 Nov
16 Nov.	Harwich		London	16 Nov
26 Nov.	London		Ipswich	27 Nov
1 Dec.	Ipswich	Lt	Yarmouth	2 Dec
12 Dec.	Yarmouth		Vlaardening	12 Dec
17 Dec.	Vlaardening	Lt	Harwich	17 Dec

The passages to and from Harwich 20th August / 1st September 1898 are made Light and it may be assumed that this period included a holiday when Jim Stone sailed CENTAUR in the Harwich Match

1928 P.R.O. ref. BT99/3942
Master E. Cripps & Mate H. Piper

1930 P.R.O. ref. BT99/4039
Master E.Cripps & Mate C.Sheldrick

1928

Date	From	Lt	To	Date
13 Jan.	London		Colchester	14 Jan
25 Jan.	Colchester		London	31 Jan
8 Feb	London		Felixstowe	10 Feb
14 Feb	Felixstowe	Lt	London	15 Feb
29 Feb	London		Colchester	1 Mar
6 Mar	Colchester		London	8 Mar
13 Mar	London		Felixstowe	15 Mar
20 Mar	Felixstowe	Lt	Colchester	21 Mar
23 Mar	Colchester		London	28 Mar
5 April	London		Ipswich	6 April
11 April	Ipswich	Lt	London	13 Aprl
8 May	London		Colchester	9 May
12 May	Colchester	Lt	London	13 May
21 May	London		Felixstowe	22 May
24 May	Felixstowe	Lt	Orford	24 May
31 May	Orford		Ipswich	31 May
2 June	Ipswich	Lt	St.Osyth	3 June
5 June	St.Osyth		London	6 June
20 June	London		Felixstowe	21 June
29 June	Felixstowe	Lt	Colchester	30 June
31 July	Colchester		London	1 Aug
20 Aug.	London		Ipswich	21 Aug
24 Aug.	Ipswich	Lt	Brightlingsea	25 Aug
27 Aug.	Brightlingsea		London	28 Aug
5 Sep	London		Felixstowe	6 Sept.
7 Sept	Felixstowe	Lt	Colchester	7 Sept
11 Sept	Colchester		London	12 Sept.
9 Oct	London		Colchester	11 Oct
13 Oct	Colchester	Lt	St.Osyth	13 Oct
17 Oct	St.Osyth		London	19 Oct
31 Oct	London		Colchester	5 Nov
14 Nov.	Colchester		London	21 Nov!
6 Dec	London		Colchester	9 Dec
13 Dec.	Colchester		London	14 Dec
22 Dec.	London		Colchester	21 Dec
30 Dec	Colchester		London	31 Dec

1930

Date	From	Lt	To	Date
15 Jan.	London		Maldon	16 Jan
20 Jan.	Maldon	Lt.	London	22 Jan
25 Jan.	London		Felixstowe	19 Jan
30 Jan.	Felixstowe	Lt.	Brightlingsea	30 Jan
1 Feb	Brightlingsea		London	3 Feb
22 Feb	London		Felixstowe	25 Feb
28 Feb	Felixstowe		London	1 Mar.
25 Mar	London		Rochford	27 Mar.
29 Mar	Rochford	Lt.	Brightlingsea	30 Mar
2 April	Britlgsea		London	2 April
2 May	London		Colchester	3 May
4 Jun	Colchester		Brightlingsea	8 Jun
2 July	London		Colchester	3 July
8 July	Colchester	Lt.	London	9 July
15 July	London		Colchester	16 July
19 July	Colchester	Lt.	London	20 July
27 July	London		Felixstowe	23 July
28 July	Felixstowe	Lt.	Brightlingsea	29 July
31 July	Brightlingsea		London	2 Aug.
13 Aug	London		Colchester	16 Aug
23 Aug	Colchester		London	26 Aug
6 Sept	London		Colchester	7 Sep
12 Sept	Colchester		London	13 Sep
22 Sept.	London	Lt.	Colchester	23 Sep
25 Sept.	Colchester		London	27 Sep
11 Oct.	London		Felixstowe	12 Oct.
16 Oct.	Felixstowe	Lt.	St.Osyth	18 Oct.
21 Oct.	St.Osyth		London	23 Oct.
4 Nov.	London		W.Mersea	5 Nov.
8 Nov.	W. Mersea	Lt.	St.Osyth	8 Nov.
10 Nov.	St.Osyth		London	12 Nov.
6 Dec.	London		Colchester	1 Dec.
16 Dec.	Colchester	Lt.	London	18 Dec.
22 Dec.	London		Colchester	25 Dec

Note Passages marked "Lt." were made light

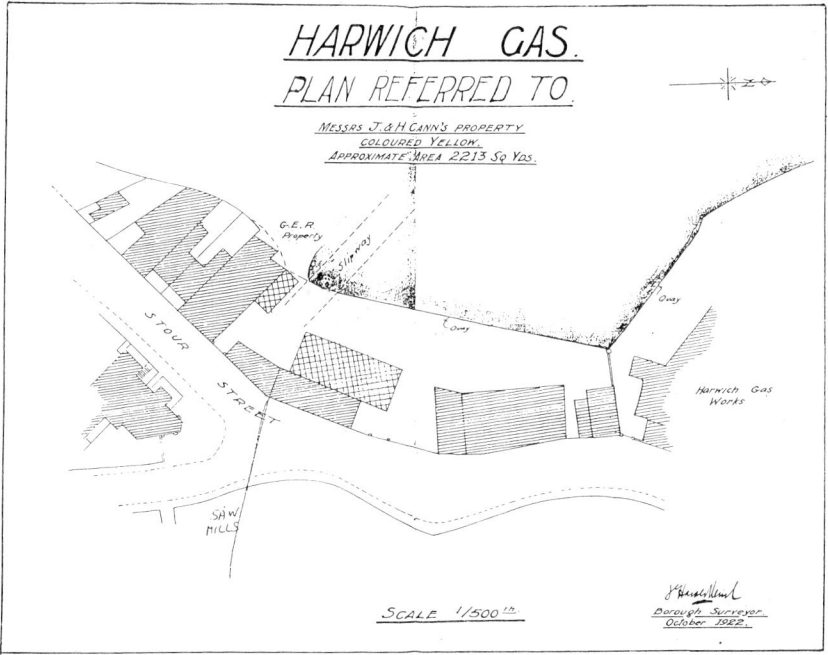

TELEPHONE Nº 5.

TELEGRAPHIC ADDRESS.
CANN, SHIPBUILDERS. HARWICH.

J & H. Cann.
Ship, Barge & Boat Builders,
Ship Chandlers, Ship & General Smiths.
Bathside, Harwich,_____19__

HARWICH GAS.
PLAN REFERRED TO.

MESSRS J.& H. CANN'S PROPERTY.
COLOURED YELLOW.
APPROXIMATE AREA 2213 SQ YDS.

G.E.R.
Property

Quay

Harwich Gas
Works

STOUR STREET

SAW
MILLS

SCALE 1/500 th.

Borough Surveyor.
October 1922.

Company Notepaper, plan of J. & H.Cann shipyard and name plate brass rubbing
Mrs. Ruth Clarke collection